Cultures of Prediction in Atmospheric and Climate Science

In recent decades, science has experienced a revolutionary shift. The development and extensive application of computer modelling and simulation has transformed the knowledge-making practices of scientific fields as diverse as astro-physics, genetics, robotics and demography. This epistemic transformation has brought with it a simultaneous heightening of political relevance and a renewal of international policy agendas, raising crucial questions about the nature and application of simulation knowledges throughout public policy.

Through a diverse range of case studies, spanning over a century of theoretical and practical developments in the atmospheric and environmental sciences, this book argues that computer modelling and simulation have substantially changed scientific and cultural practices and shaped the emergence of novel 'cultures of prediction.'

Making an innovative, interdisciplinary contribution to understanding the impact of computer modelling on research practice, institutional configurations and broader cultures, this volume will be essential reading for anyone interested in the past, present and future of climate change and the environmental sciences.

Matthias Heymann is Associate Professor for the history of science and technology at the Centre for Science Studies, Aarhus University, Denmark.

Gabriele Gramelsberger is a Professor for philosophy of digital media at the University Witten/Herdecke, Germany.

Martin Mahony is a British Academy Postdoctoral Fellow in the School of Geography, University of Nottingham.

Routledge Environmental Humanities

Series editors: Iain McCalman and Libby Robin

The *Routledge Environmental Humanities* series is an original and inspiring venture recognising that today's world agricultural and water crises, ocean pollution and resource depletion, global warming from greenhouse gases, urban sprawl, overpopulation, food insecurity and environmental justice are all *crises of culture*.

The reality of understanding and finding adaptive solutions to our present and future environmental challenges has shifted the epicenter of environmental studies away from an exclusively scientific and technological framework to one that depends on the human-focused disciplines and ideas of the humanities and allied social sciences.

We thus welcome book proposals from all humanities and social sciences disciplines for an inclusive and interdisciplinary series. We favour manuscripts aimed at an international readership and written in a lively and accessible style. The readership comprises scholars and students from the humanities and social sciences and thoughtful readers concerned about the human dimensions of environmental change.

Cultures of Prediction in Atmospheric and Climate Science

Epistemic and Cultural Shifts in
Computer-based Modelling and Simulation

**Edited by Matthias Heymann, Gabriele
Gramelsberger and Martin Mahony**

LONDON AND NEW YORK

First published 2017
by Routledge
2 Park Square, Milton Park, Abingdon, Oxon OX14 4RN

and by Routledge
711 Third Avenue, New York, NY 10017

Routledge is an imprint of the Taylor & Francis Group, an informa business

© 2017 selection and editorial matter, Matthias Heymann, Gabriele
Gramelsberger and Martin Mahony; individual chapters, the contributors

British Library Cataloguing-in-Publication Data
A catalogue record for this book is available from the British Library

Library of Congress Cataloging-in-Publication Data
A catalog record for this title has been requested

ISBN: 978-1-138-22298-4 (hbk)
ISBN: 978-1-315-40628-2 (ebk)

Typeset in Bembo
by Deanta Global Publishing Services, Chennai, India

Contents

Preface

This volume originated in conversations among a unique European research network of leading environmental scientists and historians, sociologists and philosophers of the environmental sciences, and seeks to capture some of the intensive discussions held during a series of six workshops. These discussions were initiated at an exploratory workshop in 2010 (funded by the ESF European Science Foundation) and were continued through the scientific network "Atmosphere & Algorithms" between 2010 and 2012 (funded by the DFG German Research Foundation). They have been developed further by the research group "Shaping Cultures of Prediction: Knowledge, Authority and the Construction of Climate Change" at Aarhus University between 2013 and 2017 (funded by the Danish Council for Independent Research).

We believe that among the most valuable results of these discussions were the comparative insights gained into the existence of different cultures of prediction in different scientific, institutional and political contexts. Most of our authors take a case-study approach, drawing on in-depth knowledge of the local contexts of scientific knowledge-making. Our aim has been to collect these studies together in a way which allows the emergence comparative insights across space and time; and in a fashion which we hope will inspire further research into comparative cultures of prediction.

We thank all contributors to this volume for their numerous valuable contributions to our discussions and their dedication to this project. Furthermore, we are very grateful to all other participants, contributors and local organizers of the workshops in Aarhus, Munich, Potsdam, Norwich and Paris, which are not represented in this book but were crucial for our effort. Also, numerous other colleagues, who cannot all be mentioned, deserve our gratitude, among them the team members at Aarhus University, Janet Martin Nielsen, Dania Achermann and Gabriel Henderson. The publication was supported immensely by our editor, Kelly Watkins, who was a patient and immensely helpful guide through the publishing process, and by anonymous reviewers, who helped to improve this volume with very constructive suggestions. Finally, we are grateful for the generous financial support of the European Science Foundation, the German Research Foundation and the Danish Council for Independent Research.

Matthias Heymann, Gabriele Gramelsberger, Martin Mahony

About the authors

Ralf Döscher

received a PhD in Physical Oceanography at the University of Kiel and is currently working as Science Coordinator for Global modelling activities and model development at The Rossby Center of the Swedish Meteorological Institute, Norrköping. He is the Swedish representative in European Climate Research Initiative (ECRA).

Ralf Döscher and T. Koenigk (2013) "Arctic rapid sea ice loss events in regional coupled, climate scenario experiments," *Ocean Science*, 9, pp. 217–248.

Annika E. Nilsson and Ralf Döscher (2013) "Signals from a noisy region," Miyase Christensen, A. Nilsson and N. Wormbs (eds.) *Media and the Politics of Arctic Climate Change. When the Ice Breaks*, London: Palgrave McMillan, pp. 93–113.

T. Koenigk, Ralf Döscher and G. Nikulin (2011) "Arctic future scenario experiments with a coupled regional climate model," *Tellus A*, 63, pp. 69–86.

Johann Feichter

was a senior research scientist and head of the group for "Aerosol, Clouds and Climate" at the Max Planck Institute for Meteorology, Germany. The research of Johann Feichter is focused on the development and application of numerical climate models to investigate anthropogenic influences on the climate system. Johann Feichter served as a lead author for the third and as convening author for the fourth IPCC assessment report.

Gramelsberger G. and Johan Feichter (eds., 2011) *Climate Change and Policy - The Calculability of Climate Change and the Challenge of Uncertainty*, Berlin, Heidelberg: Springer.

Johann Feichter (2011) "Shaping Reality with Algorithms: The Earth System," G. Gramelsberger (ed.) *From Science to Computational Sciences, Studies in the History of Computing and its Influence on Today's Sciences*, Zurich, Berlin: diaphanes, pp. 209–218.

Johann Feichter and P. Stier (2012) "Assessment of black carbon radiative effects in climate models ," *WIRE'S Climate Change* 3, pp. 359–370.

Gabriele Gramelsberger

is Professor for philosophy of digital media at the University Witten-Herdecke. Her research is focused on the shift from science to computational sciences, in particular driven by the introduction of computer based simulation. She has carried out an extensive study on the practice and epistemic of modelling in meteorology and cell biology during the past ten years.

Gabriele Gramelsberger, Knuuttila, T. and A. Gelfert (eds., 2013) Philosophical Perspectives on Synthetic Biology (Special Issue), *Studies in History and Philosophy of Biological and Biomedical Sciences* 44(2).

Gabriele Gramelsberger and J. Feichter (eds., 2011) *Climate Change and Policy. The Calculability of Climate Change and the Challenge of Uncertainty*, Heidelberg, Berlin, New York: Springer.

Gabriele Gramelsberger (ed., 2011) *From Science to Computational Sciences. Studies in the History of Computing and its Influence on Today's Sciences*, Zurich, Berlin: diaphanes.

Hélène Guillemot

holds a PhD in history of science, and is a researcher at Centre Alexandre Koyré (CNRS) in Paris. Her research focuses on the climate science communities and scientific practices in climate modelling. She has been working also on the relationship between science and politics and the debate and controversies about climate change.

Hélène Guillemot (2014) "Comprendre le climat pour le prévoir? Sur quelques débats, stratégies et pratiques de climatologues modélisateurs," F. Varenne and M. Silberstein (eds.) *Modéliser et simuler. Epistémologies et Pratiques de la Modélisation et de la Simulation, Tome 2*, Paris: Editions Matériologiques, pp. 67–99.

Aykut, S., Comby, J.-B. and Hélène Guillemot (2012) "Climate Change Controversies in French Mass Media: 1990–2010," *Journalism Studies* 13(2), pp. 157–174.

Hélène Guillemot (2010) "Connections between climate simulations and observation in climate computer modeling. Scientist's practices and bottom-up epistemology lessons," *Studies in History and Philosophy of Modern Physics* 41(3), pp. 242–252.

Matthias Heymann

is Associate Professor for the history of science and technology at the Centre for Science Studies, Aarhus University, Denmark. His current research focuses on the history of atmospheric and environmental sciences. He currently leads the project "Shaping Cultures of Prediction: Knowledge, Authority, and the Construction of Climate Change" and is Associate Editor of Centaurus and Domain Editor of WIREs Climate Change for the Domain Climate, History, Society, Culture.

Matthias Heymann and Janet Martin-Nielsen (eds. 2013), *Perspectives on Cold War Science in Small European States*, special issue, *Centaurus* 53(3), pp. 221–357.

Matthias Heymann (2012) "Constructing Evidence and Trust: How Did Climate Scientists' Confidence in Their Models and Simulations Emerge?" Kirsten Hastrup, Martin Skrydstrup (eds.): *The Social Life of Climate Change Models: Anticipating Nature*, New York: Routledge, pp. 203–224.

Matthias Heymann (2010) "The evolution of climate ideas and knowledge," *Wiley Interdisciplinary Reviews Climate Change* 1(3), pp. 581–597.

Nils Randlev Hundebøl

holds a PhD in Science Studies from Aarhus University. His research is focused on the diversity and development of climate science during the second half of the twentieth century as competing research groups from academia, government and industry attempted to shape and influence national and international political agendas related to the issue of climate change.

Nils Randlev Hundebøl and K.H. Nielsen (2015) "Preparing for Change: Acid Rain, Climate Change, and the Electric Power Research Institute (EPRI)," *History and Technology* 31(2), pp. 133–159.

Nils Randlev Hundebø (2014) "Challenging Climate Change: Diversity and Development in Climate Science, 1970–1995," PhD thesis, Department of Mathematics, Science Studies, University of Arhus, pp. 197.

Nils Randlev Hundebøl and K. H. Nielsen (2014) "Challenges and Social Learning at the Climate Science-Policy Interface: The 'Long Strange Trip' of the Model Evaluation Consortium for Climate Assessment (MECCA)," *Historical Studies in the Natural Sciences* 44(5), pp. 435–469.

Matthijs Kouw

is a researcher at the Rathenau Institute (The Hague, Netherlands), where he works on science-policy interfacing and technology assessment of "smart" innovations, for example, smart farming and smart cities. Matthijs holds an MA in Philosophy and an MSc in Science and Technology Studies (cum laude) from the University of Amsterdam. In 2012, Matthijs successfully defended his PhD thesis on the potential dangers of modelling practice in the domains of hydrology, hydrodynamics, soil mechanics, and ecology at Maastricht University. After completing his PhD, Matthijs was employed as a postdoctoral researcher at PBL Netherlands Environmental Assessment Agency, where he worked on climate change and values and expectations surrounding smart cities.

Matthijs Kouw (2016) "Standing on the Shoulders of Giants – and Then Looking the Other Way? Epistemic Opacity, Immersion, and Modeling in Hydraulic Engineering," *Perspectives on Science* 24(2), pp. 206–227.

Matthijs Kouw (2014) "Designing Communication: Politics and Practices of Participatory Water Quality Governance," *International Journal of Water Governance* 2(4), pp. 37–52.

Matthijs Kouw and S. van Tuinen (2015) "Blinded by Science? Speculative Realism and Speculative Constructivism," A. Longo and S. De Sanctis (eds.) *Breaking the Spell*, Milan: Mimesis International, pp. 115–30.

Catharina Landström

holds a PhD in Theory of Science from Gothenburg University in Sweden and she is currently a Senior Researcher in the School of Geography and the Environment at University of Oxford. She has worked in the United Kingdom since 2007, specializing in Environmental Science and Technology Studies (STS). Her research focuses on computer simulation modelling in relation to environmental governance and local public engagement.

Catharina Landström, R. Hauxwell-Baldwin, I. Lorenzoni and T. Rogers-Hayden (2015): "The (Mis)understanding of Scientific Uncertainty? How Experts View Policy-Makers, the Media and Publics," *Science as Culture* 24(3), pp. 276–298.

Catharina Landström and S.J. Whatmore (2014) "Virtually expert: Modes of environmental computer simulation modeling," *Science in Context* 27(4), pp. 579–603.

Catharina Landström and A. Bergmans (2014) "Long-term repository governance: a socio-technical challenge," *Journal of Risk Research* 18(3), pp. 378–391.

Martin Mahony

holds a PhD in Human Geography and Science & Technology Studies (STS) and is a British Academy Postdoctoral Fellow and Nottingham Research Fellow at the School of Geography at the University of Nottingham. His research is focused on the political history of the atmospheric sciences and on the role of assessment, simulation and visualisation at the science-policy interface. Following research on the Intergovernmental Panel on Climate Change (IPCC), he is now conducting studies on the emergence of climate modelling in the UK and on the history of meteorology as a colonial science.

Martin Mahony (2015) "Climate change and the geographies of objectivity: the case of the IPCC 's burning embers diagram," *Transactions of the Institute of British Geographers* 40, pp. 153–167.

Martin Mahony (2014) "The predictive state: Science, territory and the future of the Indian climate," *Social Studies of Science* 44(1), pp. 109–133.

Martin Mahony and M. Hulme (2012) "Model migrations: mobility and boundary crossings in regional climate prediction," *Transactions of the Institute of British Geographers* 37(2), pp. 197–211.

Janet Martin-Nielsen

holds a PhD in the history of science from the University of Toronto's Institute for the History and Philosophy of Science and Technology. She is a postdoctoral fellow at Aarhus University's Center for Science Studies. Her current research focuses on the history of climate prediction in the UK through the long twentieth century. She.has also published on the history of Arctic exploration, the history of science policy and diplomacy, and the history of linguistics.

Janet Martin-Nielsen (2015) "Re-conceptualizing the North: A Historiographic Discussion," *Journal of Northern Studies* 9(1), pp. 51–68.

Janet Martin-Nielsen (2013) *Eismitte in the Scientific Imagination: Knowledge and Politics at the Center of Greenland*, New York: Palgrave Macmillan.

Janet Martin-Nielsen (2013) "The Deepest and Most Rewarding Hole Ever Drilled," *Ice Cores and the Cold War in Greenland, Annals of Science* 70, pp. 47–70.

Annika E. Nilsson

received a PhD in environmental science and is currently working as Senior Research Fellow at Stockholm Environment Institute and Affiliated Faculty in Environmental Politics at KTH Royal Institute of Technology. Her work focuses on Arctic change, with research on environmental governance and communication at the science-policy interface. Her current research includes the project "Mistra Arctic Sustainable Development - New Governance" and "Arctic Governance and the Question of Fit in a Globalized World". She is also engaged in various Arctic Council assessments.

D. Avango, Annika E. Nilsson and P. Roberts (2013) "Assessing Arctic futures: voices, resources and governance," *The Polar Journal* 3(2), pp. 431–446.

M. Christensen, Annika E. Nilsson and Nina Wormbs (eds. 2013) *Media and the Politics of Arctic Climate Change. When the Ice Breaks*, London: Palgrave McMillan.

Annika E. Nilsson (2012) "The Arctic Environment – From Low to High Politics," L. Heininen (ed.) *Arctic Yearbook 2012*, Akureyri, Iceland: Northern Research Forum, pp. 179–193.

Markus Quante

is a senior research scientist and deputy head of the Department for Chemistry Transport Modelling at the Institute of Coastal Research of the Helmholtz-Zentrum Geesthacht, Germany. He is Professor for environmental sciences and climate physics at the University of Luneburg. His research is focused on the influence of atmospheric processes on the fate, namely the dispersion, transformation and deposition, of substances released into the atmosphere. Currently he is scientific coordinator of an international initiative conducting a comprehensive climate change assessment for the entire North Sea region.

Markus Quante, F. Colijn and the NOSCCA Author Team (eds. 2016) *North Sea Region Climate Change Assessment*, Berlin, Heidelberg: Springer.

Markus Quante, R. Ebinghaus and G. Flöser (eds. 2011) *Persistent Pollution - Past, Present, Future*, Berlin: Springer.

Markus Quante, A. Aulinger and V. Matthias (2011): "Der Schifftransport und sein Beitrag zum Klimawandel," J. Lozán, H. Graßl, L. Karbe et al. (eds.) *Warnsignal Klima: Die Meere – Änderungen und Risiken*, Hamburg: Wissenschaftliche Auswertungen, pp. 286–293.

Christoph Rosol

is research scholar at the Max-Planck-Institute for the History of Science and research associate at the Haus der Kulturen der Welt (both Berlin). His research is concerned with the epistemic foundations and technical means by which atmospheric and climate sciences have become an antetype computational science. Currently his focus is on paleoclimatology and its particular ways of entangling proxy data from the geo-archive with computer simulations to reconstruct past climates in order to calibrate outlooks into an Anthropocene future.

Christoph Rosol (2015) "Hauling Data. Anthropocene Analogues, Paleoceanography and Missing," Special Issue: Climate and Beyond, *Paradigm Shifts, Historical Social Research* 40(2), pp. 37–66.

Klingan, K., Sepahvand, A., Christoph Rosol and B. Scherer (eds.; 2014) *Textures of the Anthropocene: Grain, Vapor, Ray*, (4 vols.), Cambridge, London: MIT Press.

Christoph Rosol (2010) "From Radar to Reader. On the Origin of RFID," *The Journal of Media Geography* 5, pp. 37–49.

Birgit Schneider

is Professor for media ecology at University of Potsdam, department of media studies. Her research focuses are technical and scientific images with a strong focus on questions of mediality, codes, diagrams and textility from the seventeenth century until the present. Her current research focuses on the visual communication of climate since 1800 and a genealogy of climate change visualization between science, aesthetics and politics.

H. Bredekamp, Dünkel, V. and Birgit Schneider (eds., 2015) *The Technical Image. A History of Styles in Scientific Imagery*, Chicago, IL: Chicago University Press.

Nocke, T. and Birgit Schneider (eds., 2014) *Image Politics of Climate Change. Visualizations, Imaginations, Documentations*, Bielefeld: transcript (in cooperation with Columbia Press).

Birgit Schneider (2012) "Climate Model Simulation Visualization from a Visual Studies Perspective," *WIREs Climate Change* 3(2), pp. 185–193.

Sverker Sörlin

is Professor of Environmental History in the Division of History of Science, Technology and Environment at the KTH Royal institute of Technology, Stockholm. His current research is focused on the science politics of climate change, especially in the Arctic, and on the changes in historiography related to the Anthropocene debates and the increasing influence of global change science. He has also worked on a project on the history of environmental expertise, "Expertise for the Future", based at ANU, Cambridge, and KTH Stockholm, with a conceptual analysis of 'the environment', provisionally entitled "The Environment – a History" (with Paul Warde and Libby Robin).

W. Steffen, K. Richardson, et al. and Sverker Sörlin (2015) "Sustainability: Planetary boundaries: guiding human development on a changing planet," *Science* 347(6223), pp. 736–748.

Sverker Sörlin (2014) "Circumpolar Science: Scandinavian Approaches to the Arctic and the North Atlantic, ca 1930 to 1960," *Science in Context* 27(2), pp. 275–305.

L. Robin, Sverker Sörlin and P. Warde (eds.; 2013) *The Future of Nature: Documents of Global Change*, New Haven, CT: Yale University Press.

Nina Wormbs

is Associate Professor of History of Technology at the Division of History of Science, Technology and the Environment, KTH Royal Institute of Technology. Her research and teaching have mainly concerned issues related to decisions about and introduction of new technology in the media field. She has published on conflicts about technological change in media and takes interest in how ideas on technology shape ideas of possible action.

M. Christensen, A. E. Nilsson and Nina Wormbs (eds. 2013) *Media and the Politics of Arctic Climate Change. When the Ice Breaks*, London: Palgrave McMillan.

Nina Wormbs (2011) "Technology-dependent commons: The example of frequency spectrum for broadcasting in Europe in the 1920s," *International Journal of the Commons* 5(1), pp. 92–109.

Nina Wormbs (2006) "A Nordic satellite project understood as a trans-national effort," *History & Technology* 22(2), pp. 257–275.

1 Introduction

Matthias Heymann, Gabriele Gramelsberger
and Martin Mahony

Prediction arguably pervades all aspects of our social, political, and cultural lives. All forms of social action proceed on the basis of some expectation of what comes next, of how action will induce reaction, and of how one decisive move can trigger an unfolding of events which can lead to either a more or less desirable, or perhaps undesirable, future. From planning a barbecue to mass movement political ideologies, from the timing of sowing a crop to the design of nuclear waste deposits, reckoning with each other and with nonhuman nature means reckoning with the future, whether through tacit, experiential knowledge, or through formal, mathematised expertise. Everyday dealings with other people even depend on tacit forms of prediction, as we evaluate likely outcomes and seek a stable future in our relationships. So long as we are social, we are future-orientated.

This book examines the emergence of particular modes of orientation towards the future in an arena of science and politics where prediction has gained a particular currency: global environmental change. It focuses on the historical development of the atmospheric sciences from the early twentieth century to the present day as a case study in the emergence of new cultures of prediction which, we contend, are characteristic of modern and late-modern western societies. The rise of numerical computation in the middle of the twentieth century represents a landmark in the development of new predictive techniques, and as such the contributions to this book provide rich new insights into the epistemic and cultural shifts wrought by computation both within and beyond the environmental sciences. A comprehensive history or sociology of environmental prediction would, of course, need to encompass forms of knowledge-making, which are far removed from the major centers and practices of the formal sciences (e.g. Mathews and Barnes 2016). Here, however, we focus on the scientific tools, practices and institutions which have been central to the emergence and spread of new arguments about the future of the atmosphere and the climate, and which have consequently had deep political and cultural impacts on the ways in which the future of human societies is conceived within a changing global environment. But far from simply reifying the import of particular individuals and institutions in the production and circulation of new scientific claims, we offer here historical and sociological analyses of predictive practices in settings which have so far been overlooked in the historiography of climate science (e.g. Weart 2008; Edwards 2010; Howe 2014).

Changing prediction practices

Prediction frequently means deference to the authority and expertise of others. Oracles, priests, prophets, philosophers, fortune tellers and doctors served earlier societies' predictive needs, occupying privileged social positions justified by their access of esoteric knowledge of the future. The notion of prediction has dramatically changed, however. According to historian Reinhard Koselleck, the idea of the future as a consistent temporal category and coherent space of social imagination emerged in western thought only in the seventeenth and eighteenth centuries (Koselleck 1979; Hölscher 1999). The idea of history as a continuous development process, which was created by authors of the Enlightenment such as Ephraim Lessing, Immanuel Kant, Nicolas de Condorcet and Adam Smith, made actions and events not predestined, but historically caused. It gave rise to an increased difference between spaces of experience and horizons of expectation, as Koselleck suggests, and reflected the slow transition from a primarily religious-transcendental to a genetic and secular world view. Teleological interpretations and Christian expectations of salvation were superseded by notions of an open future and the idea of progress (Nisbet 1980: 171–236). Different futures could be conceived of and influenced by human forces – an idea which inspired modern intellectuals like Marie Jean Marquis Condorcet, Auguste Comte, Adolphe Quetelet, Friedrich List, Karl Marx and Charles Peirce (Rescher 1998: 577; Hölscher 1999: 58–63, 104–113).

The rapid and deep transformations of the nineteenth century also created new tensions and anxieties and increased the demand for prevision and reassurance. Georges Minois called the nineteenth century "a century of prediction" with an "unusual abundance of popular prophecy" (Minois 1998: 619, 614). Forms of soothsaying such as astrology, cartomancy, and chiromancy served growing demands of future knowledge by an expanding middle class. Jamie Pietruska describes the period from the mid-nineteenth century to World War I in the United States as an age of "propheteering," in which a large diversity of actors, often self-taught amateurs, served the predictive needs of farmers, traders and travelers (Pietruska 2009; see also Anderson 2005; Pietruska forthcoming). The rise of capitalist enterprise, on the other hand, with its future orientation and the inherent risks of investments made forms of economic forecasting and fortune telling a profitable new business, which clever business analysts and forecasters such as Roger W. Babson, Henry Varnum Poor, John Moody and John Knowles Fitch turned into big business after the turn of the century (Friedman 2014).

Rapid industrialization, urbanization and social struggle in the nineteenth and twentieth centuries increased demands for governmental provision. Bureaucratization in administration, military and industry promoted a specialization of planning in domains such as demography, welfare policy, insurance systems, social hygiene, urban planning, agriculture and forestry. Experiences of deep crisis in the twentieth century, such as World War I and the Great Depression, further challenged nation states to take responsibility, revise laissez-faire policies and intensify state planning and intervention. The rise of Keynesianism, the New Deal in the United States, five-year plans for the national economy in the Soviet Union, economic planning in Britain, the Four-Year Plan in Nazi Germany and Jan Tinbergen's Plan van de Arbeid (Labor Plan) in the Netherlands all reflect—although in different ways—the rising role of the state in controlling, predicting and shaping the future (e.g. Alchon 1985; Gerstle and Fraser 1989; Ritschel 1997). Provision and planning required the investigation of futures, their potential driving forces and possible points and means of intervention. Such expertise was provided, for example, by commissions such as the President's Committee on Social Trends, established by US President Herbert Hoover in 1929 in the United States or by institutions such as the Central Planning Bureau, established by Tinbergen in 1945 in the Netherlands (Friedman 2014: 166–193; Griffith 1980; den Butter and Morgan 2000).

The post-World War II era carried predictive ambitions and its social roles to new extremes. It saw, as Rescher asserts, "a growing preoccupation with the future all across the board" (1998: 28). Preoccupation with the future included a burgeoning of the predictive needs of politicians, strategists, social planners and commercial entrepreneurs as well as of popular science fiction (Andersson 2012). The frequency of usage of the term "prediction" in printed sources steeply increased after 1945. Terms such as "long-term prediction" and "short-term prediction" emerged and rapidly spread (see Figure 1.1).

The RAND Corporation, a military think tank formed in 1945 to provide methods and techniques for military and strategic planning in US administrations, neatly exemplifies the predictive obsession. It formed an active group of futurologists in the early 1950s, which pioneered many predictive techniques and, in the words of one of its members, "presaged a burst of futurological activity that was part of the Zeitgeist of the time" (Rescher 1998: 28). The ambitions surpassed older ideas of social and economic forecasting and worked across "a greatly expanded range of predictive endeavors concerned with the whole gamut of issues affecting the conditions of human life" (ibid.: 28–29). Predictive efforts and "planning euphoria" served ambitions of technological innovation, military control and social progress (Hartmann and Vogel 2010: 15–16). Exuberant confidence in the power of science went far enough to fuel attempts of modeling and predicting such complex processes as human thinking (Newell and Simon 1961), nuclear war (Ghamari-Tabrizi 2005), economic cycles (Mirowski 2002), the automation of engineering (Heymann 2005), weather and climate (Harper 2008; Edwards 2010; Fleming 2016) and the world system (Meadows et al. 1972).

Thus, it is the main claim of this book that prediction has become an intrinsic part of today's society, economy, and science. The cultural role

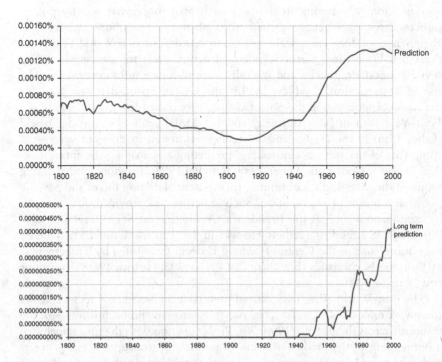

Figure 1.1 Frequencies of the terms "prediction" and "long-term prediction" in printed
sources since 1800 prepared with the Google tool N-gram Viewer. This tool
charts frequencies of any word or short sentence found in sources printed
between 1800 and 2012 in American English, British English, French, German,
Spanish, Russian, Hebrew, and Chinese. The prevalence of the term prediction
decreased during the nineteenth century and increased again after about 1915
with an accelerated increase since about 1945. The term "long-term prediction"
only turns up regularly since the late 1920s and shows a steep rise of usage in
the postwar era. Since about 1980, the use of the terms "prediction" stagnated
on a high level, whereas usage of "long-term prediction" after a short stagnation
continued to increase steeply. The n-gram for "short-term prediction" looks
very similar to the n-gram for "long term prediction".

Source: Created by the authors with Google tool N-gram Viewer.

of prediction persisted, even expanded further, when in parts of western
culture the optimistic frame of mind collapsed during the 1960s and made
way for more cautious or outspokenly pessimistic appraisals of the present
and future. New social movements questioning social conditions and institutional authorities, and growing perceptions of environmental and political
crisis, created an unsettled cultural state, which nurtured new interests in
prediction and furnished it with new meaning and import. Perceptions of
crisis in this "age of contradiction" (Brick 2000) and "fracture" (Rodgers
2011) increased the demand to fabricate coherence and certainty, stimulating

prominent engagements in the investigation of the future such as the efforts of the Club of Rome (Elichirigoity 1999; Moll 1991) and its followers and critics (Vielle Blanchard 2010). It sparked the "rise of a movement—indeed virtually an industry" of prediction (Rescher 1998: 29) led by institutions such as the Swiss think tank and consulting company PROGNOS and the newly founded International Institute for Applied Systems Analysis (IIASA). The idea and scope of prediction assumed new cultural roles and meanings. Cultural unsettledness provided rich ferment to the writings of scientists such as Barry Commoner, Rachel Carson and Paul Ehrlich, who effectively disseminated warnings about the dangers of nuclear fallout, environmental hazards and degradation and the impacts of overpopulation, and quickly became well-known public figures, featuring in numerous media interviews and television shows (Egan 2007; Sabin 2013).

These figures joined a long line of post-Enlightenment admonishers and prophets, such as Thomas Robert Malthus (1798) with his warnings about overpopulation and food security, William Stanley Jevons (1866) about coal exhaustion, and Fairfield Osborn (1948) and William Vogt (1948) on resource depletion and environmental degradation (see Robin et al. 2013; Desrochers and Hoffbauer 2009). By the 1960s, dire accounts of present conditions and future prospects were inspired and informed by the work of a new legion of scientists, particularly environmental and climate scientists, along with experts in economy, energy use, resources, demography, and other domains. Scientists took stock not only of past and present human agency and its impact on the environment, but—most importantly—engaged in systematic attempts at ever more comprehensive and increasingly long-range scientific (or science-based) prediction. Many such predictions carried the persuasive power of highly sophisticated, meticulously constructed and often diligently inspected and operated quantitative models and the hard, albeit uncertain, numbers they produced (Anderson 2010). Scientists have become, as Lynda Walsh argues, the prophets of the modern world (Walsh 2013). Prediction has become foundational and abundant in late twentieth and early twenty-first century societies, its social import so pervasive, that the very culture we live in could arguably be labelled a "culture of prediction." The modern "culture of prediction," however, is far from being investigated and understood sufficiently. This book is a modest attempt to help draw it more into the focus of scholarly investigation in the environmental humanities and history of science, to provide a frame of analysis and to engage in detailed examination of scientific, social and political practices and roles of prediction in the case of atmospheric and climate science and politics. We believe that the environmental humanities need investigations and critical appraisals of cultures of prediction, because cultures of prediction provide a key for understanding the hegemony of scientific or science-based predictive knowledge and the marginalization of humanistic and social sciences in knowledge production about human-made global environmental change—a striking disbalance which the environmental humanities seek to rectify.

Cultures of prediction

Conditions in the Cold War created the political demand and the scientific means for a radical expansion of predictions as a political means. Science was increasingly put under pressures to align with strategic, political, and social goals and to produce knowledge relevant to problem management and to technological and social innovation. Prediction as a means to advance policy became a significant part of what has been described as a new mode of knowledge production, termed variously "mode-2 science," post-normal science, or post-academic science (Gibbons et al. 1994; Funtowicz and Ravetz 1993; Ziman 2000). Digital computers, a key technological product of Cold War conditions, facilitated the treatment of complex systems, such as environmental and economic systems, and greatly broadened the reach of quantitative understanding and prediction. Computer modeling helped to represent complex systems (albeit with many simplifications) and make them amenable to scientific investigation and experimentation by controlled manipulation of the model system and the simulation of its development in space and time. Computer models, thus, helped to promote and test theoretical understanding and to surmount the restrictions of laboratory science, which cannot emulate the controlled manipulation of social systems such as the world economy and of environmental systems such as the atmosphere (Mirowski 2002; Dahan 2001). Likewise, models helped both to create and furnish social demands for predictive knowledge. It is therefore curious that the scientific and cultural import of computer modeling and simulation is yet to be subject to concerted historical and sociological inquiry, with a few landmark exceptions (Heymann 2010).

The production of knowledge about the future, fueled scientific institutions and shaped scientific cultures of their own: we suggest calling them "cultures of prediction." Sociologist Gary Alan Fine introduced the term "culture of prediction" to describe the culture of weather forecasting in the US National Weather Service's Chicago office (Fine 2007). We use the term in the plural to suggest a multitude of distinct cultures of prediction. The term "cultures of prediction" emphasizes the local origin and socially contingent character of the cultural formations built around the construction and use of computer models for predictive purposes. Cultures of prediction, like any scientific culture, operate in specific scientific and social contexts and reveal sets of shared knowledge, practices, values, and rules which emerge, stabilize, and shape scientific and public perceptions, conduct, and goals (Knorr-Cetina 1999). Compared to Fine's concept we propose a broader and at the same time more specific characterization of "cultures of prediction" by defining five key characteristics, which help to take account of the broad-ranging and pervasive role of predictive efforts in postwar modern society and make it accessible to systematic investigation. These key characteristics are:

1   the social role of prediction;
2   the character and significance of computational practices;

3 the domestication of uncertainty;
4 the degree of institutionalization and professionalization of predictive expertise;
5 the cultural impact of predictive practices and claims.

These key characteristics will be elaborated and explained in more detail in Chapter 2.

We claim that cultures of prediction based on computer modeling and simulation expanded enormously in the postwar period and have become a salient—although often invisible—feature of late twentieth- and early twenty-first-century societies with tremendous cultural and political import. The postwar period saw a change in both quantity and quality of prediction. Many social domains became intertwined and saturated with predictive efforts. Practices of prediction changed from an effort of individual authors or experts to a professionalized, collaborative business of specialized institutions. Research institutes, think tanks, private companies and governmental authorities formed part of a growing and diversifying landscape of cultures of prediction, with diverse material practices and ideological foundations. Vast numbers of scientific institutes and endeavors emerged purely for the reason of generating predictive knowledge, based often on modeling and simulation. While cultures of prediction emerged within scientific communities and built on scientific knowledge, they extend far beyond the realm of science and informed and shaped social practice, meaning and authority in broader society. Practices and meanings, which developed in local scientific communities, have migrated to new geographical locations and social groups (Jankovic 2004; Mahony and Hulme 2012; Schneider and Nocke 2014); we might say there is something imperial in the diffusion and dominance of certain predictive practices. New perceptions, visual practices and discourses have emerged and shaped the modern condition in quite fundamental ways (Turkle 2009).

The authority of cultures of prediction

The great significance of cultures of prediction lies not only in the pervasiveness, abundance and circulation of forms of predictive knowledge in societies of the late twentieth and early twenty-first centuries (Nowotny et al. 2001; Anderson 2010). Cultures of prediction represent cultures of power and, hence, transformative forces, which are all the more effective as they are often black-boxed, hidden, and invisible. Predictive knowledge has assumed a diversity of cultural roles of which three deserve particular mention: the support of politics, the justification of politics, and the replacement of politics. In the first place, cultures of prediction are generally taken to support problem solving and planning, particularly in complex environments and conditions. Many decision processes in politics, administration, or business are informed, guided, and lubricated by predictive knowledge from the planning of infrastructure to stock trading and weather prediction. Just as important as the support of decision-making is the role of cultures of prediction for the justification of decisions. Numbers generated with sophisticated scientific means such as computer models carry authority and represent powerful arguments in their own right. Models represent symbolic capital which raises the authority of the actors they serve. The implications are huge. The effectiveness of cultures of prediction in aligning perceptions and expectations and fabricating coherence contributes to the construction of broader hierarchies of knowledge and power. Cultures of prediction create dominant means of perception and understanding, framing the "distribution of the sensible," while alternative means tend to become marginalized (e.g. Hulme 2010). Dominant cultures of prediction may play a significant part in phenomena such as the marginalization of indigenous local knowledge or the marginalization of the humanities vis-à-vis predictive scientific knowledge based on computer models and simulations (Ford et al. 2016, Lövbrand et al. 2015).

As a consequence of their political, cultural, and economic status and value, tremendous resources flow towards the establishment and operation of cultures of prediction. These investments do not always serve the support or justification of decision making and politics, but can also serve to delay or replace decision making and politics—particularly in the case of contested issues with strong inherent political risks. A commonplace argument for the replacement of effective politics is the call for further research, for example, due to perceived or alleged uncertainty and the lack of sufficient knowledge apparently required to make strident decisions. Joshua Howe has made such an argument recently for the case of climate politics. While climate science provided relevant and widely-accepted predictive knowledge, the understanding of climate change did not, for a long time, translate into effective policy. Howe criticizes the "science first" approach and the reliance on science as a driver of political action. The "science first" approach served to delay politics (Howe 2013), while also creating space for climate change deniers to effectively hijack climate discourse and obstruct climate politics by promoting doubt in the results of climate science (Oreskes and Conway 2010). Others have argued that the "science first" approach invites a form of technocratic managerialism which purposefully seeks to avoid the warp and weft of agonistic, democratic political debate about the kinds of future societies want (Swyngedouw 2010; Machin 2013). In its most extreme form, this replacement of politics by science and technocracy has been advocated by actors who subscribe to the most catastrophic visions of future climate change, arguing that dire predictions demand the cessation of ordinary democratic politics in favor of a benign, pro-environmental authoritarianism, which alone can bring about the required transformation of society.

Yet, despite the apparent certainties of technocrats, predictive knowledge always involves uncertainty. Futures are uncertain by definition. Uncertainty, however, is not only and simply an epistemological problem to be worked around by scientists and policy-makers. Uncertainty, more generally, is another political resource provided by cultures of prediction, which can help to control the political role of predictive knowledge in the support, justification or replacement of politics. To become culturally effective, cultures of prediction have to domesticate uncertainty. They hedge or limit its saliency to assure the social authority of predictive knowledge and serve the support and justification of politics; and they reveal and display uncertainty to safeguard scientific authority, highlight the need for further research and serve the deferral of political action. Scientific practices such as model validation, model comparison, sensitivity studies, and expert assessments, as well as representational and rhetorical practices such as the design of graphs and the careful linguistic crafting of public statements, serve the domestication of uncertainty and production of scientific authority.

Goal of the book

This book is dedicated to the investigation of cultures of prediction in the environmental, predominantly atmospheric, sciences, which stand representative for many other domains of environmental research (as well as other disciplines and domains). It has the goal to provide answers to questions such as: What are cultures of prediction and how can they be understood, characterized, and investigated? How and why have cultures of prediction formed? How do they work and in what ways do they differ? Which typical features, seminal practices and prevalent ideologies characterize them? How is predictive knowledge about the environment produced, validated and used? How are uncertainties, inherent to any predictive knowledge, perceived, negotiated and controlled? How have cultures of prediction and the knowledge they produce gained legitimation and authority both in science and broader culture? How did scientific cultures of prediction emerge through dialogue with broader society and inform and shape social perceptions, meanings and practices?

In Chapter 2, we provide a more detailed introduction into the concept "cultures of prediction" putting the emergence of cultures of prediction based on computer modeling and simulation in broader historical perspective. Many authors claim that prediction has been a constant feature of human societies. We argue that the postwar era played a decisive role in the emergence, expansion, and dramatically increased social impact of cultures of prediction. Predictive knowledge production drew on new technologies, served new cultural interests and assumed new forms, roles, and cultural meanings. While the expansion of predictive interests represented a general tendency visible in many scientific domains, the atmospheric sciences provide a particularly powerful example for study of the features of cultures of prediction.

Part 1: Junctions: Science and politics of prediction

The chapters in the first part, titled "Junctions: Science and politics of prediction," show that the emergence and development of predictive practices was neither uniform nor uncontested, but dependent on contingent historical contexts. There was no simple and straight path to the emergence, differentiation and scientific and cultural authority of cultures of prediction. In contrast, the development of cultures of prediction involved the passing of decisive junctions, the taking of consequential decisions, the privileging of specific priorities and the neglect of potential concerns. These case studies all deal with such junctions and unveil mostly hidden competitions, struggles or conflicts between different scientific approaches and cultures in the postwar atmospheric and environmental sciences.

Gabriele Gramelsberger investigates the tradition of meteorological theory in the nineteenth century as the basis of today's weather and climate modeling, with an emphasis on meteorology in the German speaking countries. She shows that the physical understanding of atmospheric processes was a shared interest of many scientists who hoped to transform meteorology from a purely descriptive and empirical science into a quantitative physical discipline. Yet, the ninetieth century saw a struggle between various approaches to tackling the weather prediction problem and paving the way for a so-called "dynamical meteorology." Early twentieth-century meteorologists attempted to expand and apply approaches to a dynamical meteorology to create operational forms of weather prediction and also establish the foundations for a physical understanding and modeling of climate. Gramelsberger shows that the numerical approach emerged to predominance amid a variety of different scientific trajectories, fundamentally shaping twentieth-century meteorology and scientific cultures of weather prediction. This history involved radical breaks as well as intriguing continuities, not least enduring problems such as the computation of atmospheric vorticity.

The following chapter by Christoph Rosol follows this thread in the history of weather prediction and revisits the crucial years of 1945–1946, during which alternative paths to weather prediction were open for debate and discussed in the United States by Vladimir Zworykin, Vannevar Bush, and John von Neumann. Rosol shows that, in retrospect, these years proved to be formative not only for numerical modeling of atmospheric motion, but also for numerical experiments and computer simulation writ large. Zworykin, Bush, and von Neumann suggested three fundamentally different concepts for automated weather prediction, which aimed at widely different objectives oscillating between weather forecasting, weather modification, and climatic prediction. The correspondence of these central figures discloses alternative open trajectories and highlights the coproduction of theory and technology in a state of epistemic suspense.

While Rosol describes competing agendas which were hitched to the early modeling bandwagon, Janet Martin-Nielsen shows in her chapter that the emergence of modeling as a predominant research strategy was neither straightforward nor predetermined. The postwar decades saw an abundance of boundary work as discursive strategies, epistemic standards, research cultures, and claims to scientific–political–social legitimacy of new approaches to investigating climate were worked out. Martin-Nielsen investigates these contested ideas through the lens of English meteorologist and climatologist Hubert Horace Lamb (1913–1997), and his work at the UK Meteorological Office. Lamb pursued and established historical climatology as a research strategy, which conflicted with the new focus on numerical modeling in the mid to late 1960s, when new director, Basil John Mason, took office. Mason and Lamb clashed over the role of computers in weather and climate research and over the role of prediction. Martin-Nielsen sheds light on the challenges faced by traditional, empirically-minded climatologists in the postwar era, who were confronted by new actors and agendas and who were far from aligned to emerging cultures of prediction.

Matthias Heymann and Nils Hundebøl put the focus on the history of climate simulation and describe a fundamental junction: the shift from developing and using climate models as heuristic tools for improving the scientific understanding of atmospheric processes to developing and using them as predictive tools for the production of knowledge about potential future climates reaching many decades ahead. This shift occurred during the 1970s and was instrumental in the strengthening of concerns about climate warming due to increased carbon dioxide (CO_2) concentrations, in an increased visibility of climate warming as a political issue, and in raising the status of climate models from scientific tools to political instruments. This shift was hampered, however, by the uncertainty and imperfect state of climate models and was far from uncontested. Still, scientists like William Kellogg, Stephen Schneider, and James Hansen promoted the predictive use of models, because no alternative means for climate prediction and the investigation of the impact of rising CO_2 levels was apparent.

Climate modeling is an accepted research strategy as well as a contentious art. Hélène Guillemot describes some of the crucial challenges climate modeling faces, the parameterization of subgrid-scale phenomena, and explores a heated debate in the climate science community about alternative strategies of model development. Small scale processes with a spatial extension smaller than a model grid element cannot be modeled explicitly and must be approximated by so-called parameterizations, which involve significant uncertainties. Drawing from ethnographic fieldwork within a French climate modeling community, Guillemot describes the ambitious and extremely elaborate attempts to develop new cloud parameterizations based on micro-scale simulation of convection, turbulence and cloud formation. While the French group aims at improving the understanding of cloud formation and, thus, hopes to improve parameterization and climate model performance, there is no guarantee that these efforts lead to an improvement of climate prediction. Guillemot aptly calls this strategy a "gamble" and shows that it meets with considerable critique. She sheds important light on the practices, choices and visions of modelers, provides a glimpse into discussions and controversies about alternative strategies of model development and shows that epistemic conceptions of climate modeling are closely tied to scientific practices, which are themselves grounded in social and institutional frameworks.

Part 2: Challenges and debates: Negotiating and using simulation knowledge

The chapters in the second part, titled "Challenges and debates: Negotiating and using simulation knowledge," move from the junctions in the development of atmospheric and climate modeling to negotiations of uses of numerical simulation and the knowledge it produces. In these chapters, we encounter a variety of settings and practices in which the knowledge produced by simulation models has been put to work. Models and their simulation results facilitate further agency including the practices and uses of model evaluation, the dissemination and adaptation of models and modeling techniques, the coproduction of new forms of science and politics, the inspiration of contentious technological schemes such as geoengineering, and the production of new images and meanings of the future of planet Earth.

Martin Mahony explores in his chapter how regional climate modeling has emerged as a key technique since the late 1980s to inform scientists and governments about regional climatic change. Despite significant challenges caused by uncertainty and the limits of model validation, regional climate models are now in use around the world, greatly extending the reach of the major modeling centers of the global North. This chapter offers an account of the history of regional climate modeling, from the first, hesitant efforts in the western United States to the global mobility of user-friendly regional modeling tools. Exploring the institutionalization of regional modeling in the United Kingdom, Mahony argues that official regional projections, despite concerns about their reliability and utility, have been positioned as an obligatory passage point in the anticipatory governance of climatic risks. The global mobility of regional modeling and the networks coalescing around them deepen the tensions between regional climate modeling as a contested, experimental scientific practice and as a political tool in the management of risk.

Nina Wormbs, Ralf Döscher, Annika E. Nilsson, and Sverker Sörlin likewise focus on the role of climate models in the coproduction of science and politics. Their chapter reveals a paradoxical relationship between strong interest in Arctic environments, which feature as an icon of global climate change, and its long neglect in global and regional climate modeling. Investigating the conjoined scientific and political underpinnings of this paradox, Wormbs et al. show the importance of prevalent tropes of the Arctic region: for example, the idea of the Arctic as an exceptional environment, separated from the rest of the world, and of the Arctic as a unique "natural laboratory," where scientific and political interests originating largely outside of the region were pursued. In investigating how the Arctic has nonetheless been woven into discourses of "global" climate change and eventually made accessible to climate modeling, the authors show, first, that ideas of globality were rooted in local practices and places, and, second, that our understanding of responses to climate change needs to account for local forms of future imagination, which themselves emerged in scientific and political dialogue.

One of the most controversial applications of climate modeling is to the issue of geoengineering, which is being increasingly promoted as a response to the rapid environmental changes now being observed in places like the Arctic. Johann Feichter and Markus Quante pick up on the historical links between dreams of climate simulation and climate control (see also Rosol in this volume) to examine how the steady emergence of intentioned climate modification as a policy option has further transformed practices of climate modeling. Picking up on Heymann and Hundebøl's (in this volume) identification of a shift from heuristic to predictive modeling, Feichter and Quante identify a nascent shift from predictive to "instructive" modeling, whereby climate models are asked to aid in the evaluation of different geoengineering proposals. After outlining the history of geoengineering ideas and introducing the key methods, which are currently under discussion, Feichter and Quante discuss the potential of numerical climate models to explore the effects and ramifications of different geoengineering options, as well as the limitations of such model experiments designed to advise policy-makers. They suggest that this new kind of model application raises important new questions about accuracy and predictive skill. If we are to base our evaluation of controversial geoengineering methods on computer simulations, what kinds of epistemic standards should we expect these models to meet? In the context of geoengineering science we can detect the emergence of a new and potentially very consequential culture of prediction.

In Matthijs Kouw's chapter we encounter a similar conjunction of engineering, simulation uncertainties and contested epistemic standards in another highly charged political context: flood protection in the Netherlands. With sea levels expected to rise, geotechnical engineering, a sub discipline of civil engineering concerned with the behavior of soil under different conditions, fulfils an increasingly crucial function in flood protection by modeling the processes that cause flood defenses (e.g. dykes, dams, and sluices) to fail. Geotechnical modeling relies heavily on both physical models (e.g. scale models of flood defenses that are subjected to water pressures) and computational models (e.g. calculation rules that simulate the relationships between soil morphology and structural stability). Geotechnical models need to be validated to determine their ability to provide an accurate and reliable assessment of the safety of flood defenses. Drawing on insights from Science and Technology Studies (STS) and an ethnographic study of geotechnical engineering conducted at Deltares (a Dutch institute for applied research on water, subsurface and infrastructure), Kouw examines modeling practices and the validation of models pertaining to research on a dyke failure mechanism known as "piping," a form of seepage erosion. Research on piping features a series of steps and model-related forms of knowledge production, where each step produces knowledge that is made available for subsequent steps. Of particular importance are the development and adoption of computational models. However, modeling piping also features numerous uncertainties. Kouw asks how geotechnical models contribute to the production of knowledge about dyke failure mechanisms that is considered relevant by the social groups involved, and investigates the novel risks which arise through the various ways in which these social groups deal with the uncertainties involved with the use of geotechnical models.

Scientific uncertainty is an issue of perpetual interest to scientists across disciplines and, of course, to the social scientists and humanities scholars who study the creation and use of scientific knowledge. Catharina Landström's chapter considers the ways in which scientific uncertainty has been narrated in Working Group 1's (WG1) contributions to the first four Intergovernmental Panel on Climate Change (IPCC) Assessment Reports (AR). Approaching scientific uncertainty as subjected to rhetorical management, Landström traces the ways in which climate scientists have addressed it in relation to computer simulation modeling of the climate from the first IPCC AR of 1990 to the fourth in 2007, and follows online traces of uncertainty management practices from this period. The chapter explores how the emphasis of uncertainty management has shifted from reducing uncertainty by improving data, to quantifying the inevitable uncertainty arising in the process of representing the complex climate system in computer models. Landström shows how the quantitative management of uncertainty connects climate modelers with experts focusing on the mathematical analysis of model-based uncertainty, who treat uncertainty as the object of study. The interpretation of texts and web traces put forward in this chapter shows uncertainty management to be an evolving practice and suggests that quantitative analyses of uncertainty constitute a separate, emerging field of expertise.

In the final chapter, Birgit Schneider offers an analysis of the visual semantics of climate prediction, which illustrates a central theme of the book—the historic turn within climatology towards visions of the future, and all the uncertainties and ambiguities which accompany this epistemic shift. Schneider focuses particularly on how the IPCC visually frames questions of possible climatic futures. Within the semiology of multicolored maps, schemes, and charts, the IPCC condenses the most fundamental scientific understanding of climate change in its "summaries for policy-makers." About a third of the figures contained in the most recent summary for policy-makers address "the future." The figures produced for the public reader make evident how climate research today has become a producer of future knowledge, whereby climatology is expected to perform as a new kind of futurology. Taking as her starting point the perspective of visual and media studies, Schneider analyses the IPCC figures in respect to what kind of images of the future are produced scientifically (most of the images being constructed on the basis of scenarios and computer simulations), but also asks what sort of visual vocabulary is used and what colors and forms play a dominant role. The analysis leads to the question of how scientific images of possible climatic futures may be contextualized culturally. Schneider's analysis is guided by the assumption that scientific images of the future, even though they are based on matter-of-fact-numbers, statistics and simulated scenarios, are at the same time perceived within a broader cultural lineage of apocalyptic future visions, which are deeply rooted in a collective eschatological imaginary.

Together, the chapters paint a portrait of an emergent culture of prediction which pervades not only contemporary scientific practice but also the wider cultural politics of environmental change. Yet, the contributions also emphasize a diversity of cultures of prediction, showing how different scientific practices and discourses have emerged in different places, often through processes of social struggle and epistemic contestation. Despite the homogenizing practices of institutions like the IPCC, we can see in the environmental sciences how local styles of knowledge making nonetheless persist (Hulme 2008). The studies presented in this book represent new contributions to our understanding of these situated knowledge practices, and we hope that they will serve as inspiration for further research into the diversity of cultures of prediction which are both transforming science and changing the way wider societies think about, calculate, and attempt to govern the future.

References

Alchon, G. (1985) *The Invisible Hand of Planning: Capitalism, Social Science, and the State in the 1920s*, Princeton, CT: Princeton University Press.

Anderson, K. (2005) *Predicting the Weather: Victorians and the Science of Meteorology*, Chicago, IL: University of Chicago Press.

Anderson, B. (2010) "Preemption, precaution, preparedness: Anticipatory action and future geographies," *Progress in Human Geography* 34(6), pp. 777–798.

Andersson, J. (2012) "The great future debate and the struggle for the world," (AHR Forum: Histories of the future), *American Historical Review* 117(5), pp. 1411–1430.

Brick, H. (2000) *Age of Contradiction: American Thought and Culture in the 1960s*, Ithaca, NY: Cornell University Press.

Carson, R. (1968) *Silent Spring*, Boston, MA: Houghton Mifflin.

Dahan Dalmedico, A. (2001) "History and epistemology of models: Meteorology as a case study (1946–1963)," *Archive for the History of the Exact Sciences* 55, pp. 395–422.

Den Butter, F. and M. Morgan (2000) *Empirical Models and Policy-Making: Interaction and Institutions*, New York, NY: Routledge.

Desrochers, P. and C. Hoffbauer (2009) "The post war intellectual roots of the population bomb," *The Electronic Journal of Sustainable Development* 1(3), pp. 73–97.

Edwards, P. N. (1996) *The Closed World, Computers and the Politics of Discourse in Cold War America*, Cambridge, MA: MIT Press.

Edwards, P. N. (2010) *A Vast Machine: Computer Models, Climate Data, and the Politics of Global Warming*, Cambridge, MA: MIT Press.

Egan, M. (2007) *Barry Commoner and the Science of Survival, The Remaking of American Environmentalism*, Cambridge MA: MIT Press.

Ehrlich, P. R. (1968) *The Population Bomb*, New York, NY: Ballantine Books.

Elichirigoity, F. (1999) *Planet Management: Limits to Growth, Computer Simulation, and the Emergence of Global Spaces*, Evanston, IL: Northwestern University Press.

Erickson, P., J. Klein, L. Daston, R. Lemov, T. Strum and M. D. Gordin (2013) *How Reason Almost Lost its Mind, The Strange Career of Cold War Rationality*, Chicago, IL: Chicago University Press.

Fine, G. A. (2007) *Authors of the Storm. Meteorologists and the Culture of Prediction*. Chicago, London: University of Chicago Press.

Ford, J. D., L. Cameron, J. Rubis, M. Maillet, D. Nakashima, A. Consolo Willox and T. Pearce (2016) "Including indigenous knowledge and experience in IPCC assessment reports," *Nature Climate Change* 6(4), pp. 349–353.

Fleming J. R. (2016) *Inventing Atmospheric Science*, Cambridge, MA: MIT Press.

Friedman, W. A. (2014) *Fortunetellers: The Story of America's First Economic Forecasters*, Princeton, CT: Princeton University Press.

Funtowicz, S. O. and J. R. Ravetz (1993) "Science for the post-normal age," *Futures* 25(7), pp. 739–755.

Gerstle, G. and S. Fraser (eds. 1989) *The Rise and Fall of the New Deal Order*, Princeton, CT: Princeton University Press.

Ghamari-Tabrizi, S. (2005) *The Worlds of Herman Kahn: The Intuitive Science of Thermonuclear War*, Cambridge, MA: Harvard University Press.

Gibbons, M., C. Limoges, H. Nowotny, S. Schwartzman, P. Scott and M. Trow (1994) *The New Production of Knowledge*, London, UK: Sage.

Griffith, R. T. (1980) "The Netherlands Central Planning Bureau," R. T. Griffith (ed.), *The Economy and Politics of the Netherlands Since 1945*, Dordrecht, Netherlands: Springer Netherlands, pp. 135–161.

Harper, K. C. (2008) *Weather by the Numbers: The Genesis of Modern Meteorology*, Cambridge, MA: MIT Press.

Hartmann, H. and J. Vogel (2010) "Prognosen: Wissenschaftliche Praxis im öffentlichen Raum," in: H. Hartmann and J. Vogel (eds.), *Zukunftswissen: Prognosen in Wirtschaft, Politik und Gesellschaft seit 1900*, Frankfurt, Germany: Campus, pp. 7–29.

Heymann, M. (2005) *Kunst und Wissenschaft in der Technik des 20. Jahrhundert, Zur Geschichte der Konstruktionswissenschaften*, Zurich, Switzerland: Chronos.

Heymann, M. (2010) "Understanding and misunderstanding computer simulation: The case of atmospheric and climate science – An introduction," *Studies in History and Philosophy of Modern Physics* 41(3), pp. 193–200.

Hölscher, L. (1999) *Die Entdeckung der Zukunft*, Frankfurt: Fischer Taschenbuch Verlag.

Howe, J. (2014) *Behind the Curve, Science and Politics of Global Warming*, Seattle, WA: University of Washington Press.

Hulme, M. (2008) "Geographical work at the boundaries of climate change," *Transactions of the Institute of British Geographers* 33(1), pp. 5–11.

Hulme, M. (2010) "Problems with making and governing global kinds of knowledge," *Global Environmental Change-Human and Policy Dimensions* 20(4), pp. 558–564

Jankovic, V. (2004) "Science migrations: Mesoscale weather prediction from Belgrade to Washington, 1970–2000," *Social Studies of Science* 34, pp. 45–75.

Jevons, W. S. (1866) *The Coal Question*, (2nd revised edition), London: Macmillan and Co.

Knorr-Cetina, K. (1999) *Epistemic Cultures: How the Sciences Make Knowledge*, Cambridge, MA: Harvard University Press.

Koselleck, R. (1979) *Vergangene Zukunft. Zur Semantik geschichtlicher Zeiten*, Frankfurt: Suhrkamp (English translation Koselleck, R. (2004) *Futures Past. On the Semantics of Historical Time*, New York, NY: Columbia University Press).

Lövbrand, E., S. Beck, J. Chilvers, T. Forsyth, J. Hedrén, M. Hulme, R. Lidskog and E. Vasileiadou (2015) "Who speaks for the future of Earth? How critical social science can extend the conversation on the Anthropocene," *Global Environmental Change* 32, pp. 211–218.

Machin, A. (2013) *Negotiating Climate Change: Radical Democracy and the Illusion of Consensus*, London: Zed Books.

Mahony, M. and M. Hulme (2012) "Model migrations: mobility and boundary crossings in regional climate prediction," *Transactions of the Institute of British Geographers* 37(2), pp. 197–211.

Malthus, T. R. (1798) *An Essay on the Principle of Population*, London: J. Johnson

Mathews, A. S. and J. Barnes (2016) "Prognosis: Visions of environmental futures," *Journal of the Royal Anthropological Institute* (N.S.) 22(2), pp. 9–26.

Meadows, D. H., D. L. Meadows, J. Randers and W. W. Behrens III (1972) *The Limits to Growth: A Report for the Club of Rome's Project on the Predicament of Mankind*, New York, NY: Universe Books.

Minois, G. (1998) *Geschichte der Zukunft. Orakel, Prophezeiungen, Utopien, Prognosen*, Düsseldorf, Zurich, Swtizerland: Artemis & Winkler.

Mirowski, P. (2002) *Machine Dreams: Economics Becomes a Cyborg Science*, Cambridge, MA: Cambridge University Press.

Moll, P. (1991) *From Scarcity to Sustainability. Futures Studies and the Environment: The Role of the Club of Rome*, Frankfurt, Germany: Peter Lang.

Newell, A. and H. A. Simon (1961) "Computer simulation of human thinking," *Science* 134(3495), pp. 2011–2017.

Nisbet, R. (1980) *History of the Idea of Progress*, New Brunswick: Transaction Publishers.

Nowotny, H., P. Scott and M. T. Gibbons (2001) *Re-thinking Science: Knowledge and the Public in an Age of Uncertainty*, London: Blackwell.

Oreskes, N. and E. M. Conway (2014) *The Collapse of Western Civilization: A View from the Future*, New York, NY: Columbia University Press.

Osborn F. (1948) *Our Plundered Planet*, Boston, MA: Little, Brown and Company.

Pietruska, J. (2009) *Propheteering: A Cultural History of Prediction in the Guilded Age*, Dissertation, Cambridge, MA: MIT Press.

Pietruska, J. (forthcoming) *Looking Forward: Prediction and Uncertainty in Modern America*, Chicago, IL: University of Chicago Press (forthcoming).

Rescher, N. (1998) *Predicting the Future: An Introduction to the Theory of Forecasting*, New York, NY: SUNY Press.

Ritschel, D. (1997) *The Politics of Planning, The Debate on Economic Planning in Britain in the 1930s*, Oxford, UK: Oxford University Press.

Robin, L., S. Sverker and P. Warde (eds. 2013) *The Future of Nature: Documents of Global Change*, New Haven, CT: Yale University Press.

Rodgers, D. T. (2011) *Age of Fracture*, Cambridge, MA: Belknap Press.

Sabin, P. (2013) *The Bet: Paul Ehrlich, Julian Simon, and Our Gamble Over Earth's Future*, New Haven, CT: Yale University Press.

Schneider, B. and T. Nocke (eds. 2014) *Image Politics of Climate Change. Visualizations, Imaginations, Documentations*, Bielefeld: transcript.

Seefried, E. (2015) *Zukünfte. Aufstieg und Krise der Zukunftsforschung 1945–1980*, Berlin, Germany: De Gruyter.

Swingedouw, E. (2010) "Apocalypse forever? Post-political populism and the spectre of climate change," *Theory, Culture & Society* 27(2–3), pp. 213–232.

Turkle, S. (2009) *Simulation and its Discontents*, Cambridge, MA: MIT Press.

Vielle Blanchard, E. (2010) "Modelling the future: An overview of the 'limits to growth' debate," *Centaurus* 52, pp. 91–116.

Vogt W. (1948) *Road to Survival*, New York, NY: William Sloan.

Walsh, L. (2013) *Scientists as Prophets: A Rhetorical Genealogy*. Oxford, UK: Oxford University Press.

Weart, S. R. (2008) *The Discovery of Global Warming*, (2nd revised and expanded edition), Cambridge, MA: Harvard University Press.

Ziman, J. (2000) *Real Science: What It Is and What It Means*, Cambridge, MA: Cambridge University Press.

2 Key characteristics of cultures of prediction

Matthias Heymann, Gabriele Gramelsberger and Martin Mahony

"It is very hard to make predictions—especially about the future." This verdict, often attributed to Niels Bohr (Rescher, 1998: 2) and originally a popular Danish proverb, ironically casts the unattainability of a deep-rooted human desire and social need. Served through the ages by dedicated agents such as oracles, priests, and prophets in ancient worlds to cultures of prediction in modern times, the future was always part of the present, although in very different ways. An increasing corpus of literature has made the argument that in modern society, particularly in the postwar world, prediction attained hugely expanded significance (Koselleck 1979; Rescher 1998; Hölscher 1999; Hunt 2008; Andersson 2012; Andersson and Rindzevičiūtė 2015; Seefried 2015). Our aim in this chapter is to develop the concept "cultures of prediction" as a heuristic frame to study the practices and cultures of "futurework" (Fine 2007), analyze its social roles and contribute to an understanding of its pervasiveness in modern, particularly postwar societies.

What are cultures of prediction and why have they become particularly abundant and influential in the late twentieth century? This is the major question we wish to deal with in this chapter and—for the case of the environmental, atmospheric, and climate sciences—in the whole book. This concept is meant not only to provide details and clues about the scientific and social peculiarities of prediction practices, but also to reveal and understand its dramatically increased significance and pervasive role in postwar society. Cultures of prediction both reflect and shape idiosyncrasies of the modern condition. They accommodate ambitions and anxieties, serve demands and desires, shape the world we perceive and the questions we ask, and bring forth practices and products, policies, and politics. Most importantly, they construct hierarchies of knowledge and hegemonic epistemologies, which determine in many ways how our world works, looks, and feels.

The term "cultures of prediction" is not our invention. Ann Johnson and Johannes Lenhard have applied this term to describe an immense expansion of computer simulation work with the advent of the personal computer (PC). They emphasize the wide range of disciplines, domains, and individual applications of PC based simulation. Hence, they see the cheap, easily accessible, and abundant PC as the instrument which turned a specialty of a few experts—computer simulation and prediction—into a mass phenomenon and created a "new culture of prediction" (Johnson and Lenhard 2011; Mathew and Barnes 2016: 12). While Johnson and Lenhard address a large scale phenomenon in a sweeping overview, sociologist Gary Alan Fine coined the term "culture of prediction" to describe the microculture of weather prediction in his comprehensive ethnographic study of three offices of the US National Weather Service (NWS) in the mid-western United States (Fine 2007). According to Fine, forecasting is a collective action and prognoses of the future are "shaped by the contours of group life" (ibid.: 2).

In his book, Fine does not explicitly generalize characteristics of what he calls a "culture of prediction," but closely follows his protagonists in ethnographic observations to distill the features of their local culture. His interest focuses on themes such as practices and identities, the role of science and the authority it creates, the depiction of future events and the construction of predictive accuracy (which he calls "futurework"), the maintaining of occupational autonomy by control of language and images and the relation of future workers and others who stand outside the boundary of their workplace (ibid.: 3). Fine shows how weather forecasters occupy intermediary positions— for example between modeling and data analysis or between science and the public. The tensions from being pulled in multiple directions, the inherent ambiguity of their work and the local peculiarities, practices, and personalities shape their specific occupational culture (ibid.: 142, 242). His observational effort revealed meteorology as "something akin to art, a personalistic and elusive process of interpretation, a domain of authenticity that is beyond the abilities or even understanding of outsiders" (ibid.: 13).

Fine's book provides an impressive range of observations and insights. He claims that weather forecasting culture is a highly localized "idioculture," as he frames it, effective on the scale of groups, based in shared group settings and "arising from and contributing to small group dynamics" (ibid.: 69). His work raises the question of how specific local cultures transmute into or become part of a more comprehensive, cohesive epistemic culture (Knorr-Cetina 1999). How does a multitude of idiosyncratic group cultures form a larger epistemic community, which is unified by recognized competence and expertise and "a shared set of normative and principled beliefs" (Haas 1992: 1)? "Notwithstanding local differences, forecasting practice at the NWS [National Weather Service] is aligned around a common organizational process and shared core principles of meteorological decision making," Phaedra Daipha has emphasized (Daipha 2015: 56). Daipha's work, building on Fine and also based on an ethnographic study of local weather prediction practice, investigates the process of meteorological uncertainty management and decision making and provides an example of "moving up the evidential chain from a specific case to a general conceptual framework of decision making" (ibid.: 216).

Similarly, we attempt to offer a broadened framework of *cultures of prediction* to describe and better understand postwar predictive efforts based on computer simulation. This concept builds on investigating local practice and culture and an appreciation for historical, cultural, and scientific detail and, at the same time, provides an analytical frame to proceed from the specifics of idioculture to the features of a broader epistemic culture, characterized by a shared "grammar" of practice and understanding with broader social and cultural bearing (Fujimura 1992). We define the concept "cultures of prediction" with the help of five key characteristics or dimensions, which serve its detailed analysis:

1   the social role of prediction;
2   the character and significance of computational practices;
3   the domestication of uncertainty;,
4   the degree of institutionalization and professionalization of predictive expertise; and
5   the cultural impact of predictive practices and claims.

These dimensions will be explicated in more detailed below.

The social role of prediction

Prediction is always pursued for a social purpose. This claim may appear self-evident or even tautological, as every human practice has a social purpose. Still, it deserves elaboration. Daniel Sarewitz and Roger A. Pielke distinguish two different types of prediction, which they aptly labeled "predictions as a means to advance science" and "predictions as a means to advance policy" (Sarewitz and Pielke 2000: 17). The first type of prediction serves as "a test of scientific understanding … Scientific hypotheses are tested by comparing what is expected to occur with what actually occurs. When expectations coincide with events, it lends support to the power of scientific understanding to explain how things work." The second type of prediction is "a potential guide for decision making. We may seek to know the future in the belief that such knowledge will stimulate and enable beneficial action in the present" (ibid.: 11–12). Both types perform powerful social roles, whether in the social worlds of science or policy-making, but this book deals with primarily the second type of prediction. Throughout history and throughout the diversity of human societies past and present, humanity has been characterized by social needs for prediction and afforded an immense variety of predictive cultures and prediction techniques; occult, religious, philosophical, and scientific (Minois 1998). We suggest that a first dimension in the investigation of cultures of prediction is the analysis of the broad social roles of predictive efforts—whether they deal with astrological divination, religious prophecy, economic prognosis, weather forecasting or climate prediction. The task is to trace these practices and knowledges from their sites of origin to their interactions with other social groups, and their enrolment into broader social and cultural projects.

The atmospheric sciences provide a convenient example. Advancing policy and contributing to societal debates are key goals in some areas of these sciences. Weather prediction is an obvious case, where the demands of agriculture, transport, the military, and other domains have spurred the development of forecasting techniques since the nineteenth century. More recently, climate prediction has become one of the major preoccupations of both scientists and broader societies. In 1988, the Intergovernmental Panel on Climate Change (IPCC) was established by the United Nations Environment Programme (UNEP) and the World Meteorological Organization (WMO) "to provide the world with a clear scientific view on the current state of knowledge in climate change and its potential environmental and socio-economic impacts" and "rigorous and balanced scientific information to decision makers," as its homepage explains (IPCC 2016). Its social purpose was made explicit at the United Nations General Assembly in summer 1988, which called on the IPCC to "immediately initiate action leading ... to a *comprehensive* review and *recommendations* with respect to ... the science of climate change ... [and] possible policy responses by Governments to delay, limit or mitigate the impact of adverse climate change" (cited in Agrawala 1998a: 616). The Fifth Assessment Report of the IPCC, published in 2013, presented climate projections based on simulations with 42 climate models, predicting a warming of between one and four degree Celsius by the year 2100 (IPCC 2013).

Prediction for a social purpose, however, often comes at a price. Venturing into the large-scale effort of climate prediction and the production of policy-relevant knowledge was a significant feat of atmospheric scientists. It promised significant social rewards (funding and reputation, to say the least) and helped to make climate modeling and prediction a global "big science" (Schützenmeister 2009). More recently, however, some climate scientists have started to question the value of the immense effort put into the preparation of climate projections for the IPCC. This work keeps thousands of scientists and many of the largest computers worldwide busy, but—as they argue—does not promise substantial improvement of the understanding of the atmosphere, which has been stagnating for a number of years (Held 2005; Bony et al. 2011; see also Heymann and Hundebøl in this volume). The case of climate modeling shows that it makes a difference to scientists whether they use their expertise and technologies to advance scientific understanding or whether they focus on social purposes such as climate prediction. These different goals can give rise to very different institutional formations, as Shackley (2001) shows in his study of climate modeling centers (see also Krueck and Borchers 1999). Pursuing prediction for a social purpose requires clear-cut decisions. It not only channels resources in a certain direction (funding, workforce, computer time in the case of climate modeling); it also impacts the very models that scientists use, because these have to be made fit to the specific task in question (Shackley 2001).

Prediction can be a contentious and contested effort, often balancing across a thin line between the perception that enough knowledge and understanding is available for it and the apprehension that knowledge and means for the scope of prediction is insufficient, prediction not reliable enough. Where this boundary is to be set is a matter of social negotiation and dependent on cultural factors and values. This boundary, however, is soft and porous. Strong social demand and attractive rewards for prediction may constitute a significant incentive to provide predictions, even if its basis is considered weak (better a bad prediction than no prediction). In the last decades of the nineteenth century, weather prophets who predicted the development of seasonal weather found a whole regiment of customers, even though the US Weather Bureau kept cautioning against weather prophecies of quacks and charlatans (Pietruska 2009). For the soothsayers and augurs in premodern societies to consultants in the late twentieth century, prediction certainly was a profitable business, often regardless of whether predictions turned out to be right or not (Rescher 1998: 13; Randalls 2010). Examining the particular social roles of prediction across diverse domains is an important starting point for understanding the shape and prevalence of modern cultures of prediction. By taking a historical perspective on domains such as the atmospheric and environmental sciences, as many of the following chapters do, we can begin to piece together not only changing practices of prediction and foresight over time, but the effects of such knowledge claims on the organization, politics and cultures of broader social worlds.

The character and significance of computational practices

As a second dimension, the analysis of the character and significance of computational practices is essential for understanding cultures of prediction. Such practices may appear insignificant to a wider public, to which only the end product matters (e.g. a weather prediction). But these practices structure and constrain human perception and legitimate action in significant ways. The doing, crafting, and toiling of scientists engaged in forging predictive claims, the material culture these practices constitute, have a bearing on the ways scientists see, comprehend and describe the world (Knorr-Cetina 1999). Andrew Pickering has called this the "mangle of practice," which mutually forms material culture, human perception, and scientific claims (Pickering 1995). Therefore, the chapters of the first part of this volume concentrate on the development of computational practices. They investigate the crucial role of predictive practices making use of apparatus, devices, machines, or systems, which "script" and regulate routines, configure predictive ideas and approaches, and put constraints on perception and vocabulary. Latour has famously argued that machines have agency in the creation of scientific data (Latour 1987). Fine acknowledges the role of technology and emphasizes the importance of the human-machine relationship in weather forecasting. He argues that the machine and the forecaster represent a "heterogeneous couple" with the machine ritualizing scientific practice and shaping the meaning of our world. "Equipment tames and transforms the world, but by this we are tamed and transformed" (Fine 2007: 105).

The most influential predictive efforts in the twentieth century involved computation, whether in domains of the natural sciences (such as weather forecasting) or in more social domains (such as economic prognosis). Quantitative information based on scientific procedures often carry superior authority. The foundational role of computation in the forging of predictive claims explains the importance of means of computation, from human computers and mechanical calculators such as adding machines to the programmable computer of the postwar era. The rise of computing capacity in the twentieth century through the mechanization of computing fueled an immense growth of computational practice, among others, for the purpose of prediction (Grier 2007). The programmable computer of the postwar era with exponentially rising computational power sparked a true computational revolution. British scientist, Lewis Fry Richardson, who pioneered quantitative weather prediction derived from physical equations, estimated in 1922 that the computation of a weather forecast for the next day would require a workforce of 64,000 human computers (Richardson 1922: 219). Such effort was beyond feasibility—until the digital computer arrived a few decades later.

Leading domains in the atmospheric sciences, in fact, exemplify the explosion of computational needs. Computation is so much at the core of things that atmospheric scientists use the biggest computers, need incredible amounts of computer time—and still complain about the limitations of computer power (Heymann 2010a). The atmosphere is a huge and complex physical entity, consisting of different gases and governed by a large number of physical and chemical processes on different temporal and spatial scales. In 1904, Norwegian physicist Vilhelm Bjerknes laid out the basis for a solution to the problem of weather prediction. He suggested seven non-linear partial differential equations, the so-called "primitive equations," which described in principle all atmospheric processes and meteorological states at any point in time and space (Bjerknes 1904; Thorpe et al. 2002; Gramelsberger 2009; see also Gramelsberger in this volume). Yet, these equations could not be solved analytically. During World War I, Lewis Fry Richardson attempted an approximate, so-called "numerical solution" of Bjerknes' primitive equations, the computation of approximate average solutions on a spatial grid dividing central Europe into a grid of 23 squares. Although this grid was extremely coarse compared to today's resolutions the computations proved so tedious and lengthy—they took Richardson six full weeks for a one-day weather prediction on two grid elements—that no weather prediction could realistically be accomplished before the weather to be predicted had already passed (Richardson 1922; Lynch 2006). It was not until World War II and the Cold War that military interests and funding paved the way for numerical weather prediction on one of the first digital computers. In 1950, using drastically simplified versions of Bjerknes' primitive equations, a team around mathematician John von Neumann and meteorologist Jule Gregory Charney computed for the first time a weather prediction along the lines of Richardson's approach (Harper 2008; Nebeker 1995; see also Rosol in this volume). Thanks to the computer (and a group of enthusiastic and brilliant scientists) weather simulation was born.

This breakthrough was revolutionary, because it made for the first time a complex system like the atmosphere amenable to quantitative mathematical investigation. Norman Phillips, a member of von Neumann and Charney's team, simulated a forecast period of 30 days with a further simplified version of the weather model. While this experiment was not based on realistic initial conditions, it turned out to be surprisingly successful in reproducing patterns of the atmospheric circulation. Phillips concluded that "the verisimilitude of the forecast flow patterns suggests quite strongly that it [the model] contains a fair element of truth" (Phillips 1956: 154). The experiment was path-breaking in two ways: first, it showed that computer based simulation could serve to simulate atmospheric phenomena; second, it proved that "[n]umerical integration of this kind ... give[s] us [the] unique opportunity to study largescale meteorology as an experimental science," as the British meteorologist, Eric Eady, concluded in 1956 (quoted in Lewis 1998: 52). If the atmosphere could not be investigated experimentally in the laboratory, experiments could be pursued with a model on a computer. The computer became the laboratory (Gramelsberger 2010).

Atmospheric modeling brought about a host of practices, largely invisible and inaccessible to the outside world. These practices included:

- the manipulation of physical equations, computer model construction and the invention of radical simplifications on many levels;
- the manufacture and manipulation of massive amounts of data as input for model simulation;
- the massive production of simulated data, which soon outnumbered observation based data;
- virtual simulation experiments to test concepts, models and assumptions;
- model validation strategies for comparing simulation results with observation-based data; and
- sensitivity studies to test the sensitivity of models to changed (uncertain or not reliably known) data and assumptions.

Computer modeling and simulation involve a great deal of trial and error, model engineering, troubleshooting, debugging, testing, and tuning (Heymann, 2010a). This is also true for the processing and uses of modeling results. Phaedra Daipha describes weather forecasting as a "culture of disciplined improvisation." The practice of weather prediction represents a form of "collage," "a process of assembling, appropriating, superimposing, juxtaposing, and blurring information" (Daipha 2015: 15, 21). These practices help to create credible models and to manufacture sufficient coherence between characteristics of the observed and virtual worlds, data and models, model building and simulation purpose, observation and simulation, past experience and constructed futures, scientific values and public expectations, uncertainty and reliability. These practices structure scientific perceptions and shape a broader culture of weather prediction.

Climate modeling and prediction, though based on many similar techniques, represent a very different culture. While weather forecasters are expected to predict the weather for the next days as accurately as possible, climate modelers, in contrast, produce projections of future climate change based on a range of emission scenarios. Weather forecasters' simulations mainly depend on a large range of atmospheric input data. Climate modelers need to base their simulations on atmospheric data and on socio-economic and political scenarios, which in themselves are sources of great uncertainty. Weather modelers routinely test their models every day by comparing simulation and actual weather data. Climate modelers do not have that luxury, because the only empirical testbeds they have available are data on past climates.

These modeling practices do not just entail certain habitual sequences of action, but are loaded with ideas, ideals, and ideologies. Deeply ingrained in postwar simulation practice was a positivistic confidence in the possibility of scientific prediction. This ideology emerged in the nineteenth century, when scientists and scholars of cosmic, geological, and biological evolution, from Immanuel Kant and Jean Baptist Lamarck to Thomas Malthus, Herbert Spencer, Charles Darwin and William Wallace, expressed new predictive ambitions (Rescher 1998: 26). In particular, theorists such as Pierre Louis Maupertuis and Pierre-Simon Laplace emphasized the authority and pivotal role of scientific prediction. In his *Philosophical Essay on Probabilities*, first published in French in 1812, Laplace developed the idea later famously referred to as "Laplace's demon." An intelligence which could on the basis of physical principles comprehend all natural forces and the state of all things would know all "movements by the greatest bodies of the universe and those of the lightest atom; for it, nothing would be uncertain and the future, as the past, would be present to its eyes" (Laplace 1902: 4). Science seemed a proper means to allow control over nature and plan and shape the future. Even the impossibility of knowing things fully could be compensated for, if only imperfectly, Laplace suggested: "We owe to the weakness of the human mind one of the most delicate and ingenious of mathematical theories, the science of chance or probabilities" (cited in Gillispie 1972: 6). He regarded probability theory as a viable approach to generating predictive knowledge and finding answers to both scientific problems such as celestial mechanics and political problems such as demography.

The basis of predictive authority changed not only the scientific expertise represented in specialized data and knowledge, sophisticated technology and expert practice, but also the understanding of the objects under investigation. Thus, in particular, climate modeling did not simply expand knowledge about climate, but shaped a new perception and understanding of climate and created new interests in and images of climate (Heymann 2009, 2010c; see also Heymann and Hundebøl in this volume). As Chris Russill (2016) has argued, the emergence of a geophysical approach to understanding climate, shaped by general circulation models, sidelined alternative approaches which emphasized local ecologies and climate-society interactions, extremes rather than global means, and risk management frameworks for dealing with climatic change. Global climate models, with their new emphasis on global trends, could speak more meaningfully to economic models whose creators and users were likewise concerned with the rational management of trends. Here again, the mechanics of particular kinds of model shaped what it was possible to meaningfully say about a phenomenon, and how it was possible to intervene in it.

The domestication of uncertainty

Any prediction necessarily involves uncertainty. Uncertainty potentially devalues predictive claims and divests them of their persuasive power and social effectiveness. Weather forecasting, Daipha writes, "is arguably the most iconic example of a decision-making task riddled with deep uncertainty" (Daipha 2015: 2). In comparison, economic prognosis appears even more troubled with uncertainty. Likewise, climate modelers are haunted by the uncertainties of data, models, and simulation results (Gramelsberger and Feichter 2011). How do producers of predictive knowledge domesticate uncertainty to stabilize their claims and make them trustful and effective in society and politics? The investigation of the domestication of uncertainty is a third dimension in the analysis of cultures of prediction and many chapters of part two of this volume focus on this issue (see chapters by Kouw, Landström and Schneider).

In the ideology of modern science, uncertainty is perceived as an unavoidable corollary of scientific work—all scientific knowledge is preliminary, defective and incomplete—and uncertainty is a quality of knowledge which scientists seek to control and reduce (Nowotny 2016). Objective criteria for defining "good science" are held to be available, which include procedures of testing knowledge claims and keeping uncertainty at bay (Shackley et al. 1998). The domestication of uncertainty, however, is a more foundational category than these commonplace statements suggest. All human existence is fundamentally uncertain. Predictive efforts are one way to contain this uncertainty and alleviate the contingencies and risks of decision making and of mastering everyday life. Uncertainty in this anthropological sense is a condition of life and prediction a response to it. Lynda Walsh argues that "polities call on science advisers to manufacture certainty for them, not uncertainty" (2013: 196). Fine concurs: "The dark heart of prediction is defining, controlling, and presenting uncertainty as confident knowledge" (2007: 103). The question is how can cultures of prediction accomplished this? There hardly is a general answer to this question. The perception of what constitutes sufficiently certain knowledge varies across research domains, cultural contexts and the interests of diverse actors.

Climate modelers risk the immediate loss of political and public trust and, potentially, even their jobs, if their results turn out to be false and their models and practices unreliable. Forecasts about the development of economic indicators such as growth, investments and unemployment, on the other hand, fail fairly regularly without seriously compromising the status of the economists (Mirowski 2013). Domesticating uncertainty is further complicated by the fact that uncertainty represents a resource that is mobilized for different ends by scientists, regulators, industry, media and other actors. Uncertainty can furnish powerful arguments for regulation to protect safety and health in the face of risks, or conversely to reject regulation due to the lack of certain knowledge (Jasanoff 1990; Oreskes and Conway 2010). Hence, domesticating uncertainty is not simply a matter of inventing practices to produce robust and reliable knowledge. It is a matter of conflict, negotiation, and boundary work and is intricately linked to the establishment of social credibility, legitimacy, and authority of scientific claims and policy responses. Scientific credibility may be negotiated within a community of experts, but it hinges on how claims of expertise fare in the outside world (Mahony 2014a; Gieryn 1999). Domesticating uncertainty is a social process involving all actors invested in prediction, the scientists producing and framing it, the customers interpreting and making use of it, competitors questioning it, interest groups challenging it, media distorting it, and so on around the cultural circuits of predictive knowledge (Lahsen 2005; Novotny 2016).

Since Edward Lorenz's recognition in the 1960s of the inherent unpredictability of non-linear systems like the climate (Fleming 2016, Heymann and Hundebøl in this volume), atmospheric modelers have been continually haunted by many sources and types of uncertainty such as a lack of knowledge about atmospheric processes, a dearth (or sometimes a glut) of data, drastic simplifications in the representation of processes in models, limited spatial and temporal resolution and limits of computer power. These (and other) problems often require pragmatic approaches, sometimes even ingenious inventions and work-around solutions, which are accepted and justified if model validation is deemed successful and a good fit is obtained with empirically observable processes. Atmospheric modelers usually posit model validation as the litmus test of model performance. But what is a sufficiently successful validation and what is not?

A model that performs well and accomplishes good fits between calculated and measured values is usually seen as an appropriate "verification of the reference model" (Eliassen et al. 1982: 1646). However, this covers the complexity of the relation between measurement data and model data and the diverse practices of measurement and prediction (a detailed case analysis of pollution transport modeling is presented in Heymann 2006). There are many problems of evaluation and various forms of uncertainties involved, questioning the possibility of validation (or verification) of model results – instead the term evaluation is used (Oreskes et al. 1994; Petersen 2006). Furthermore, evaluation is a social process based on shared practices, norms, and values and on agreement and compromise. Fine called it an "interactional achievement" (2007: 16). For the case of climate modeling, Hélène Guillemot has shown that there is no general protocol for the evaluation of climate models. Norms of validation have been shaped locally and differ in different institutions and cultural contexts (Guillemot 2010).

In the case of weather forecasts, Jamie Pietruska reports a conflict between the US Weather Bureau and the self-taught weather prophet W. T. Foster in the late nineteenth century. Foster claimed successful weekly, monthly, and yearly weather forecasts, which earned him a reasonable income. The US Weather Bureau at the same time fashioned what Pietruska calls a "culture of predictive certainty," conducting systematic evaluations to assess its forecasters' accuracy and maintain the public reputation of an authoritative government weather service. According to the Weather Bureau the legitimate range of a scientific forecast was two to three days in summer and one to two days in winter. Any claims of long-term prediction of weather were regarded as charlatanry. Consequently, the Weather Bureau investigated an evaluation chart for a monthly prediction provided by Foster in order to prove the fraudulent character of his claims. Not surprisingly, Willis M. Moore, the head of the Weather Bureau in Washington, declared "practically no agreement" between the predicted and recorded temperatures "except a few coincidental verifications." Foster, instead, claimed 66 to 75 percent accuracy of prediction and questioned the Bureau's standards for validation. He contended that a "variation of one day, sometimes two, from the forecast will be accounted good by those most interested and is so claimed in my publications. A difference of a few degrees between the forecast and the temperature does not vitiate the forecast in the minds of those who are watching them closely" (Pietruska forthcoming: 186–188). The Weather Bureau had a hard time to establish and preserve a status of uncontested authority. It engaged in boundary work to establish and maintain a demarcation between scientific forecasting (Weather Bureau) and fraud (Foster) to preserve its primacy and superiority in weather prediction (for different cases of such boundary work see Henry 2015; Martin-Nielsen in this volume). The Bureau's efforts to domesticate uncertainty by cultivating a "culture of predictive certainty" and rejecting long-term prediction altogether, was not necessarily more successful than Foster's strategy to contain uncertainty by allowing "some variation" and a "difference of a few degrees" between forecast and actual weather. His customers depended not on exact predictions, but on the general development of the weather, whether it arrived one day or the next. As Lahsen (2005) has shown in a more recent context, perceptions of the uncertainty associated with a particular knowledge claim, and the standards by which it should be judged, vary greatly between producers and users of predictive tools and knowledges.

Domesticating uncertainty may prove an easier task, if predictive cultures operate invisibly, either because nobody cares, or because the predictions only silently serve limited groups of customers (for example consultants serving the management of corporations). Predictive efforts may also be protected behind walls of technocratic expertise and become accepted and black-boxed routines (for example in specialized agencies). Weather and climate prediction, in contrast, work under conditions of a public science and continuously have to invest in domesticating uncertainty and in impression management to maintain scientific and public authority. Communication from the exact wording of texts to the design of figures and graphs becomes a crucial part of domesticating uncertainty and launching scientific claims in the public sphere (Schneider and Nocke 2014; Mahony 2015). The more contested knowledge claims and public authority are the more must be invested in the containment of uncertainty and in rhetorical devices and strategies. Public and political challenges to the claims of climate prediction pushed the IPCC to carefully define procedures and protocols, devise terminologies for the treatment of uncertainty and create a virtual science of uncertainty (see also Landström in this volume).

The degree of institutionalization and professionalization of predictive expertise

A fourth dimension in the analysis of cultures of prediction is the investigation of the degree of institutionalization and professionalization of predictive expertise. While many prophets, soothsayers and fortunetellers worked alone, by the twentieth century this type of fortune telling remained a thriving commercial enterprise (Minois 1998: 712–727). At the same time, politically and culturally significant efforts at prediction became increasingly sophisticated and collaborative, characterized by a growing degree of institutionalization and professionalization.

Many predictive efforts such as weather and climate prediction or economic forecasting depended on large amounts of data provided through observation networks or statistical surveying and bookkeeping, and on sophisticated technology and technical expertise. Nineteenth-century weather offices were often highly differentiated institutions running observational networks and many local offices, employing wireless telegraphy and teams of telegraphers for the exchange of weather information, and hosting a workforce of weather experts for analysis, map drawing, weather prediction and public communication (Fleming 1990; Anderson 2005). Business analysts and economic forecasters such as Roger W. Babson also built huge, comprehensive forecasting companies in the early twentieth century. Babson established the Babson Statistical Organization. He hired "teams of statisticians" and, as Friedman described it, "created something of a kingdom." In 1920, his company had a production department, a marketing department, and an administrative department and employed a staff of about 300 (Friedman 2014: 15, 38). By the late twentieth century, the Intergovernmental Panel on Climate Change was involving thousands of scientists for the assessment of knowledge about future climate change within a comprehensive framework of rules defining status, structure and procedures (Agrawala 1998b; Bolin 2009).

According to Fine, forecasting is a form of collective action based on "a system of distributed knowledge" (2007: 28, 32; see also Mahony in this volume). Institutionalization provides structures of formalized and routinized operation, defines rules and uniformity of practice, and determines interaction and collaborative work. The institution builds a protected space and provides a roof under which a diversity of resources such as specialized knowledge and expertise, sophisticated technology and operational protocols can be gathered, along with experience and tradition, an ethos of apprenticeship, and shared vocabularies, values, standards and norms (e.g. Mahony and

Hulme 2016). Institutions of prediction resemble what Latour called "centres of calculation" (Latour 1987: 215–257), which are characterized by a systematic mobilization of human ("actors") and non-human resources ("actants"). They form a network and "mutually negotiate their roles, attributes and competencies in the process of knowledge production until a well-functioning web of allies is stabilized that makes up a new knowledge claim" (Jöns 2011: 159). Forecasting institutions act like centers of calculation in constructing their claims. In some cases, they mirror scientific research institutions and forecasting remains a form of "science in action." Often, however, as in the case of weather prediction, the institutionalization of forecasting represents a move away from traditional forms of scientific work to standardized operational practice, in which prediction becomes a routine guided by defined procedures (Fine 2007; Daipha 2015).

The internal structure, social role, and historical dynamics of institutionalized forecasting, such as weather prediction, could likewise be described with Thomas P. Hughes' concept of a socio-technical system. Formed around a crucial technology, these systems develop stability and momentum with increasing complexity and size (Hughes 1983, 1989). Predictive practices and institutions are built around technologies, for example, technologies of observation, compilation, computation, analysis, and presentation. While the incandescent lamp served as a starting point for the establishment of powerful electricity systems, computer models and simulation based on the primitive equations and computer technology helped to build a powerful socio-technical system of weather prediction. The metaphor of the technological system emphasizes order, persistence, growth, and inertia of predictive institutions. According to Fine, technology also plays a role in furnishing predictive expertise and institutions with authority and creating "institutional legitimation" of predictive claims (Fine 2007: 101). "Technology doesn't just produce data, it validates it and makes it worth paying for. Equipment justifies the scientist," Fine argues, and more broadly the institution (ibid.: 106). Technologies define the specialized knowledge and competence that is required for the predictive effort and, at the same time, constitute a barrier of access to authoritative expertise to keep non-experts out.

In her ethnographic study of weather prediction at the US NWS, Phaedra Daipha presents a strong example of the role of technology in disciplining practice, constructing authority and contributing to institution building. In the 1990s, the NWS introduced a new Advanced Weather Interactive Processing System (AWIPS), which became the "nerve centre of NWS forecasting operations" and "truly anchored the NWS forecasting routine onto the computer" (Daipha 2015: 34). In the early 2000s, the NWS introduced another component of AWIPS, the Interactive Forecast Preparation System (IFPS). IFPS was a software system which "radically transformed the NWS forecasting routine and propelled NWS operations into the digital age of weather forecasting" (ibid.). It changed forecasting from being based on geographical regions to being based on a uniform grid system, and from being text based to graphics based—"text forecast is now automatically generated from the graphical forecast" (ibid.: 35). All forecasting information had to be generated with a Graphical Forecast Editor and was sent to the central NWS server, the so-called "National Digital Forecast Database." The system constrained the freedom of the individual forecaster significantly and, on the other hand, aligned procedures and styles in forecasting offices nationwide. The radical disciplining of practice caused enormous resistance and forms of subversion in the initial years after the introduction (ibid.: 43–51). On the other hand, AWIPS and IFPS increased the detail of forecasting information and services and served to strengthen the institution and its authority. According to Daipha, the IFPS shifted "symbolic and material emphasis away from expert judgment and toward mechanical objectivity" (ibid.: 54). Furthermore, the NWS made its Digital Database an "obligatory passage point" to render itself indispensable. "Without a doubt, the digitization of the NWS forecasting routine embodies an anxious attempt on the part of the NWS to remain viable and relevant as a government agency ... in an increasingly competitive weather market" (ibid.: 35).

Professionalization and institutionalization of forecasting often developed hand in hand and, historically, have shown a tendency to increase in degree. Fortunetellers in the late nineteenth century emphasized the scientific basis of their techniques, often referring to their unique "system" of forecasting for services such as the estimation of future crops, weather or economic development. Many weather prophets employed self-constructed "systems" for long-range forecasting, which were based on the periodicities of weather data, planetary meteorology, lunar phases, animal behavior, plant growth or other indicators. Regular appearance in newspapers or popularity through widely circulated almanacs, made them authoritative public institutions, although mostly on a local scale, sometimes even competing with and challenging the institutional authority of the Weather Bureau (Pietruska 2009: 138–176).

In the thriving field of economic analysis and forecasting, Roger W. Babson started his business with the "Babson Card System," a device designed to provide information on index cards about bonds offered by different companies. Later, he made the collection of business information the basis for analyzing future trends in a weekly newsletter "Babson's Reports" and in book form titled "Business Barometers." His system of information collection and manipulation was the basis for building the Babson Statistical Organization as a strong institution during the 1920s (Friedman 2014: 18–39). Similarly, Tinbergen's "schemata," "mathematical machines" or "models," as he called them became the core device or apparatus around which he formed the Central Planning Bureau in 1945, an influential government body responsible for economic forecasting and economic policy analysis (Maas 2014: 48; Griffith 1980; den Butter and Morgan 2000).

While the prewar era already brought about the cultural and scientific conditions for a rise of model-based prediction, only the postwar era saw its rapid expansion, institutionalization and exploitation based on computer technology. Powerful new institutions emerged such as the RAND Corporation (founded in 1945), PROGNOS AG (founded in 1959) and the International Institute for Applied System Analysis (IIASA, founded in 1972). In addition, major organizations such as the International Monetary Fund (IMF, founded in 1944), the Organization for Economic Co-Operation and Development (OECD, founded in 1961) as well as governmental authorities, research institutions and major banks and corporations established modeling and forecasting departments (Schmelzer 2016; Seefried 2015; den Butter and Morgan 2000). Economic forecasting also remained a profitable private business. Specialized institutions such as the Wharton Econometric Forecasting Associates (WEFA), the Data Resources, Inc., and Chase Econometrics (all founded in 1969) became "great cathedrals of econometrics, developed massive forecasting models and sold their predictions" (Friedman 2014: 206). The institutionalization of prediction in the postwar era is so abundant, but at the same time mostly so invisible, black-boxed and hidden that its historical impact and significance has been hard to assess. This volume is one contribution to a growing field of scholarship which is aiming to do just that.

The cultural impact of predictive practices and claims

A fifth dimension in the analysis of cultures of prediction is the investigation of the broader cultural impact of predictive practices and claims. How and why are cultures of prediction effective and how and why do they inform and shape the culture they are part of? (See also Wormbs et al. in this volume; Feichter and Quante in this volume; Schneider in this volume) The discursive preoccupation with climate change is a notable example. Despite being largely invisible to human senses, the processes of climate change, and projections of their future course, have become parts of our cultural life in ways in which other serious threats—such as air pollution or possible energy shortages—have not. A climate-changed future has come to feature prominently in political, economic, religious, ethical, and artistic discourses about how we relate to each other and to the nonhuman world (Hulme 2009), sparking new ways of comprehending futurity through concepts such as intergenerational justice, tipping points, and civilizational collapse (e.g. Oreskes and Conway 2014). Such concepts tie together the knowledges and modes of anticipation of various disciplines and modes of thought, constructing new visions of futures which challenge deep-seated notions of justice, responsibility, and progress.

The cultural traction of climate change was perhaps prefigured by the transformative effects of the publication of *The Limits to Growth*, the 1972 Report to the Club of Rome, (Meadows et al. 1972). A research group around Dennis L. Meadows working under Jay W. Forrester at the Massachusetts Institute of Technology made use of Forrester's System Dynamics approach, developed to investigate complex systems, such as urban and industrial dynamics, with the aid of computer simulation (Elichirigoity 1999: 40–59). In the early 1970s, he built two versions of a mathematical so-called "world model," named "World 1" and "World 2" for computer simulations of future global developments (Forrester 1971). Meadows and his group built an adjusted model called "World 3" to simulate and examine the development of five variables—world population, industrialization, pollution, food production, and resource depletion. The simulations showed that unabated population growth and resource use would lead to stagnation and collapse "within the next century, at the latest" (Meadows et al. 1972: 126). The purpose of *The Limits to Growth* was to explore how exponential growth interacts with environmental conditions and finite resources. Launched with a highly effective marketing campaign, ultimately translated into 30 languages and selling over 10 million copies worldwide by 1999, it boosted the idea of environmental and resource crisis and of an inherent danger of societal collapse sometime in the twenty-first century. Despite much criticism (Taylor and Buttel 1992), the cultural impact of this report was truly "monumental" on Western societies (Elichirigoity 1999: 3, 5). This was a piece of work which was a product of an era of proliferating environmental concern, concerns which it furnished with new ammunition—not least the apparent ability to peer into the future of complex, interacting systems. It thus raises the question of how wider cultural concerns drive and shape predictive activity, and how prediction in turn re-shapes wider societal discourses of environmental anxiety.

Both of these cases of prediction, contemporary climate science and the *Limits to Growth* report, represent a globalization of global environmental concern—a globalization in the sense that the object of concern is definitively the "global" environment, and also in the sense of globalizing the cultural impact of predictive practices. We might ask whether prediction in the post-war era reached new heights of cultural saliency, as some authors suggest (e.g. Rescher 1998) and others question (e.g. Minois 1998). Lynda Walsh traces a continuous line linking the cultural roles of prophets from the Pythia, known as the oracle of Delphi, to Francis Bacon, Robert Oppenheimer, Carl Sagan, and the IPCC. Studying the words of contemporary scientists, she noticed "how similar their speech patterns were to the biblical prophecy" and how similarly they worked in ancient as in contemporary society. In her rhetorical genealogy of prophetical ethos, she argues that scientists have persistently assumed the cultural practices constituting prophetical ethos by constructing a "persistent, recognizable cluster of rhetorical strategies". Scientists, she contends, have served as the prophets of the modern world, empowered and authorized to manufacture "political certainty" (Walsh 2013: iix, 2).

The cultural impact of prediction, on the other hand, does not only depend on actors engaged in prophesy and their rhetorical tools, but on scientific and technological means, institutional settings and cultural and political contexts. Only the development of the digital computer during and after World War II made Bjerknes' equations tractable and numerical weather prediction feasible. Only the digital computer opened a door for the enormous venture of climate and earth system modeling and prediction. Only the digital computer helped to accelerate calculation and experimentation with Tinbergen's econometric models effectively, to develop massive new economic forecasting models able to process huge amounts of data within a very short time and to facilitate a quick expansion of cheap and widespread economic simulation in numerous institutions. But digital computing was itself the product of particular political cultures and contexts, not least the Cold War and the renewed emphasis placed on surveillance, planning, prediction, and control. In this context, computers became significant new tools, but also prominent cultural icons in the shaping of new understandings of human subjectivity, artificial intelligence, and complex systems (Edwards 1996). The digital computer did not create the new enthusiasm for planning and prediction, but it amply supported ideas and ambitions to pursue these goals and, at the same time, created new perceptions of how it was possible act on and intervene in the world.

Similarly, the scientific paradigm change from data- and statistics-based pattern recognition and extrapolation to quantitative mathematical modeling based on mathematical equations, represented a fundamentally different set of possibilities with deep consequences. Mathematical equations established clear causal relationships, and mathematical modeling a way of thinking in deterministic (and reductionist) causal relations. Modeling, first, established a new way of describing and understanding systems such as weather and economic cycles. Second, modeling encouraged the idea that targeted interventions in such systems is possible. Modeling represented a toolbox for thriving ambitions of environmental and social control, be it weather and climate control or the control of economic cycles. Third, modeling placed the parameter of time at center stage and focused attention on the details of temporal change, whereas details of change across space faded into the background. National and global business cycles and global climate change are prominent examples. Contemporary society is excited and bothered about (global and large-scale) temporal change as it is established and prognosticated by economic, climate, and other models. Furthermore, modeling efforts have favored a focus on the large scale while arguably marginalizing the small and the local, which has always proved much harder to reliably and convincingly represent in mathematical models with global ambitions.

For at least two reasons the investigation of the cultural impacts of predictive practices and claims is not only crucial for the understanding of cultures of prediction, but for an understanding of society at large: transcendence and power. The cultural effectiveness of cultures of prediction is hardly visible and far from obvious, because predictive claims and the understandings, values and norms they shape and carry have become self-evident and normalized ways of experiencing the world. We can hardly see these cultures of prediction, not only because many of them are hidden behind the walls of expert institutions, but because they define how we see and what we see. They transcend perceptibility. At least, we don't see them *as* cultures of prediction doing something with the world. Numerical weather prediction has become so ingrained in our ever-day routines that many of us forget just how profoundly its omnipresence shapes culture. It is hard to conceive what a world without numerical weather prediction would look like. Climate prediction has created a new understanding of climate as a global entity changing in time, an understanding which did not exist in the mid-twentieth century when scientists began to experiment with computer models (Miller 2004). Investors, producers, traders, and decision makers' understandings and life-worlds are critically shaped by models and graphs plotting the change of entities over time, from the past and far into the future. The question of performativity, of the worlds these epistemic constructs bring into being, is fundamental for understanding the salience of cultures of prediction.

Power is a second and related issue. As cultures of prediction shape the world we see and experience, they have considerable leverage on the distribution of epistemic power and to some extent define which actors are influential and which are not (Ashley 1983). The weather prophet, W. T. Foster, possessed significant power. His forecasts satisfied a cultural demand. They were covered by many newspapers and shaped the decisions of farmers, traders, and others. His forecasting system and public presence rendered sufficient authority to challenge even the Weather Bureau. Roger W. Babson established a whole empire of economic forecasting and became a public celebrity at the peak of his business, with agencies to sell the Babson Reports in London, Paris, and Berlin (Friedman 2014: 32). His influence derived not least from reducing the complexity of economic processes to index numbers and capitalizing on thinking in economic cycles in ways compatible with the comprehension and demands of decision makers. The Club of Rome commanded tremendous influence with the publicizing of the *Limits to Growth* report, channeling hitherto unseen attention worldwide to the question of future economic and environmental collapse.

Power is not just influence, as a generation of work on the intimate relationship, or even of co-constitution, of knowledge and power makes clear (e.g. Foucault 1980; Rouse 1987; Jasanoff 2004). Cultures of prediction create discourses which shape practices, subjectivities, symbolic capital, and hierarchies of knowledge. Atmospheric and climate knowledge in the late twentieth century was dominated by the physical sciences and physics-based modeling approaches and reasoning. This knowledge has been co-produced with new forms of global political imagination and ambition (Miller 2004), in ways which make some "solutions" to the problem of climate change (such as carbon trading) more legible, more tractable, and thus more politically convenient, than others (e.g. Lövbrand et al. 2009). On the level of states, the authority to make legitimate contributions to climate knowledge and discourse in the twenty-first century seems to require the operation of a national climate modeling and forecasting effort, leading to the global proliferation of earth system models (Mahony 2014b; Mahony and Hulme 2016). National modeling efforts are arguably not just about providing useful, "local" information to decision-makers, but about the performance of governmental competence and epistemic sovereignty for both national and global audiences (see also Mahony in this volume). Furthermore, certain standardized climate models seem to have become an "obligatory passage point" when it comes to authoritative policy formulation in widely different regions and cultural settings (Mahony and Hulme 2012). On the level of disciplines, geographers, with their attention to spatial variation and local detail, found themselves marginalized from new debates about global climate change in the 1980s, even though they had largely defined the paradigm and identity of climatology until only few decades ago (Hulme 2008; Heymann 2009, 2010c; Russill 2016). Similarly, perspectives from the social sciences and humanities as well as forms of local and indigenous knowledge about climate have been superseded by the hegemony of science-based modeling (Bjurström and Polk 2011). Dissenting voices have recently become louder, but still have a hard time making themselves heard and reclaiming some intellectual ownership of the idea of climate and of climate-society interactions (Ford et al. 2016; Lövbrand et al. 2015).

Theodore M. Porter argued that the emerging interest in quantification and statistical procedures in the nineteenth century derived from the need of decision makers to construct legitimation. Traditional face-to-face dealings lost their importance and were replaced by longer chains of interactions and dependency. Belief in the self-evident trustworthiness of traditional elites declined. In turn, quantitative and procedural forms of accountability became increasingly important. Porter calls them "technologies of distance" (Porter 1995: ix). In response to pressures from outside, "strategies of impersonality" created authority, particularly in settings where trust was in short supply. Technologies of distance and strategies of impersonality helped to make uncertain knowledge "take the form of objectivity claims" and "not depend too much on the particular individuals who author it" (ibid.: 229). Quantification and statistics, it seems today, were only first steps: modeling and simulation have since transformed the toolbox of exercising power and changing the world. The promises of modeling, among them the promise of prediction, have been sources of both epistemic and political power, fundamentally re-shaping expectations about the links between nature and society, science and politics, and technology and culture. Understanding the nature and workings of this power is a key task for students of cultures of prediction, requiring all the intellectual tools of the critical social sciences and the environmental humanities.

References

Agrawala, S. (1998a) "Context and early origins of the Intergovernmental Panel on Climate Change," *Climatic Change* 39, pp. 605–620.

Agrawala, S. (1998b) "Structural and process history of the Intergovernmental Panel on Climate Change," *Climatic Change* 39, pp. 621–642.

Andersson, J. (2012) "The great future debate and the struggle for the world, AHR Forum: Histories of the future," *American Historical Review* 117(5), pp. 1411–1430.

Andersson, J. and E. Rindzevičiūtė (eds. 2015) *The Struggle for the Long-Term in Transnational Science and Politics. Forging the Future*, New York, NY: Routledge.

Ashley, R. (1983) "The eye of power: The politics of world modeling," *International Organization* 37(3), pp. 495–535.

Bjerknes, V. (1904) "Das problem der wettervorhersage, betrachtet vom standpunkt der mechanik und physik," *Meteorologische Zeitschrift* 21, pp. 1–7.

Bjurström, A. and M. Polk (2011) "Physical and economic bias in climate change research: A scientometric study of IPCC Third Assessment Report," *Climatic Change* 108, pp. 1–22.

Bolin, B. R. J. (2009) *A History of the Science and Politics of Climate Change: The Role of the Intergovernmental Panel on Climate Change*, Cambridge, MA: Cambridge University Press.

Bonneuil, C. and J.-B. Fressoz (2016) *The Shock of the Anthropocene*, London, UK: Verso.

Bony, S., B. Stevens, I. H. Held, J. F. Mitchell, J.-L. Dufresne, K. A. Emanuel, P. Friedlingstein, S. Griffies and C. Senior (2011) "Carbon Dioxide and Climate: Perspectives on a Scientific Assessment," *WCRP Open Science Conference website*, http://www.wcrp-climate.org/conference2011/documents/LongTermClimateChange_Bony.pdf.

Boumans, M. (2005) *How Economists Model the World into Numbers*, New York, NY: Routledge.

Daipha, P. (2015) *Masters of Uncertainty. Weather Forecasters and the Quest for Ground Truth*, Chicago, IL: University of Chicago Press.

Edwards, P. N. (1996). *The Closed World: Computers and the Politics of Discourse in Cold War America*, Cambridge, MA: MIT Press.

Edwards, P. N. (2010) *A Vast Machine: Computer Models, Climate Data, and the Politics of Global Warming*, Cambridge, MA: MIT Press.

Eliassen, A., J. Saltbones, F. Stordal, Ø. Hov and I. S. A. Isaksen (1982) "A Lagrangian long-range transport model with atmospheric boundary layer chemistry," *Journal of Applied Meteorology* 21, pp. 1645–1661.

Elichirigoity, F. (1999) *Planet Management: Limits to Growth, Computer Simulation, and the Emergence of Global Spaces*, Evanston, IL: Northwestern University Press.

Fine, G. A. (2007) *Authors of the Storm. Meteorologists and the Culture of Prediction*, Chicago, IL: University of Chicago Press.

Fleming J. R. (1990) *Meteorology in America, 1800–1870*, Baltimore, MD: John Hopkins University Press.

Ford, J. D., L. Cameron, J. Rubis, M. Maillet, D. Nakashima, A. Cunsolo Willox and T. Pearce (2016) "Including indigenous knowledge and experience in IPCC assessment reports," *Nature Climate Change* 6(4), pp. 349–353.

Forrester, J. W. (1971) *World Dynamics*, Cambridge, MA: Wright-Allen Press.

Foucault, M. (1980). *Power/Knowledge: Selected Interviews and Other Writings*. London, UK: Harvester Press.

Frängsmyr, T., J. Heilbron and R. E. Rider (eds. 1990) *The Quantifying Spirit in the Eighteenth Century*, Berkeley, CA: University of California Press.

Friedman, W. A. (2014) *Fortunetellers: The Story of America's first Economic Forecasters*, Princeton, NJ: Princeton University Press.

Fujimura, J. H. (1992) "Crafting science: Standardized packages, boundary objects, and 'translation'," A. Pickering (ed.) *Science as Practice and Culture*, Chicago, IL: University of Chicago Press, pp. 168–211.

Gieryn, T. F. (1999) *Cultural Boudaries of Science: Credibility on the Line*, Chicago, IL: Chicago University Press.

Gillispie, C. C. (1972) "Probability and politics: Laplace, Condorcet, and Turgot," *Proceedings of the American Philosophical Society* 116, pp. 1–20.

Gramelsberger, G. (2009) "Conceiving Meteorology as the exact science of the atmosphere: Vilhelm Bjerknes's paper of 1904 as a milestone," *Meteorologische Zeitschrift* 18, pp. 669–673.

Gramelsberger, G. (2010) *Computerexperimente. Zum Wandel der Wissenschaft im Zeitalter des Computers*, Bielefeld: Transcript.

Gramelsberger, G. and J. Feichter (eds. 2011) *Climate Change and Policy. The Calculability of Climate Change and the Challenge of Uncertainty*, Heidelberg, Germany: Springer.

Grier, D. A. (2007) *When Computers were Human*, Princeton, NJ: Princeton University Press.

Guillemot, H. (2010) "Connections between simulations and observation in climate computer modelling, Scientist's practices and 'bottom-up epistemology' lessons," *Studies in History and Philosophy of Modern Physics* 41, pp. 242–252.

Haas, P. M. (1992) "Introduction. Epistemic communities and international policy coordination," *International Organization* 46(2), pp. 1–35.

Hacking, I. (1990) *The Taming of Chance*, Cambridge, MA: Cambridge University Press.

Harper, K. C. (2008) *Weather by the Numbers: The Genesis of Modern Meteorology*, Cambridge, MA: MIT Press.

Held, I. (2005) "The Gap between simulation and understanding in climate modeling," *Bulletin of the American Meteorological Society* 86, pp. 1609–1614.

Henry, M. (2015) "'Inspired divination': Mapping the boundaries of meteorological credibility in New Zealand, 1920–1939," *Journal of Historical Geography* 50, pp. 66–75.

Heymann, M. (2005) *Kunst und Wissenschaft in der Technik des 20. Jahrhundert. Zur Geschichte der Konstruktionswissenschaften*, Zurich, Switzerland: Chronos.

Heymann, M. (2006) "Modeling reality: Practice, knowledge, and uncertainty in atmospheric transport simulation," *Historical Studies of the Physical and Biological Sciences* 37, pp. 49–85.

Heymann, M. (2009) "Klimakonstruktionen. Von der klassischen Klimatologie zur Klimaforschung," *NTM. Journal of the History of Science, Technology and Medicine* 17(2), pp. 171–197.

Heymann, M. (2010a) "Understanding and misunderstanding computer simulation: The case of atmospheric and climate science – An introduction," *Studies in History and Philosophy of Modern Physics* 41(3), pp. 193–200.

Heymann, M. (2010b) "Lumping, testing, tuning: The invention of an artificial chemistry in atmospheric transport modeling," *Studies in History and Philosophy of Modern Physics* 41(3), pp. 218–232.

Heymann, M. (2010c) "The evolution of climate ideas and knowledge," *Wiley Interdisciplinary Reviews Climate Change* 1(3), pp. 581–597.

Heymann, M. (2012) "Constructing evidence and trust: How did climate scientists' confidence in their models and simulations emerge?" K. Hastrup and M. Skydstrup (eds.) *The Social Life of Climate Change Models: Anticipating Nature*, New York, NY: Routledge, pp. 203–224.

Horn, E. (2014) *Zukunft als Katastrophe*, Frankfurt, Germany: Fischer Taschenbuch Verlag.

Hughes, T. P. (1983) *Networks of Power: Electrification in Western Society, 1880–1930*, Baltimore, MD: Johns Hopkins University Press.

Hughes, T. P. (1989) "The evolution of large technological systems," W. Bijker, T. Hughes, and T. Pinch (eds.) *The Social Construction of Technological Systems*, Cambridge, MA: MIT Press, pp. 51–87.

Hulme, M. (2008) "Geographical work at the boundaries of climate change," *Transactions of the Institute of British Geographers* 33(1), pp. 5–11.

Hunt, L. (2008) *Measuring Time, Making History*, Budapest, Hungary: Central European University Press.

IPCC (2013) *Climate Change 2013. The Physical Science Basis*, Cambridge, MA: Cambridge University Press.

IPCC (2016) *Homepage*, available from http://www.ipcc.ch/organization/organization. shtml, accessed July 2016.

Jasanoff, S. (1990) *The Fifth Branch, Science Advisers as Policy Makers*, Cambridge MA: Harvard University Press.

Jasanoff, S. (ed. 2004). *States of Knowledge: The Co-Production of Science and Social Order*, London, UK: Routledge.

Johnson, A. and J. Lenhard (2011) "Toward a new culture of prediction: Computational modeling in the era of desktop computing," A. Nordmann, H. Radder and G. Schiemann (eds.) *Science transformed? Debating Claims of an Epochal Break*, Pittsburgh, PA: University of Pittsburgh Press, pp. 189–199.

Jöns, H. (2011) "Centres of calculation," J. Agnew and D. N. Livingstone (eds.) *The SAGE Handbook of Geographical Knowledge*, London, UK: Sage Publications, pp. 158–170.

Kern, S. (2003) *The Culture of Time and Space 1880–1918*, Cambridge, MA: Harvard University Press.

Knorr-Cetina, K. (1999) *Epistemic Cultures: How the Sciences Make Knowledge*, Cambridge, MA: Harvard University Press.

Krueck, C. and J. Borchers (1999) "Science in politics: A comparison of climate modelling centres," *Minerva* 37(2), pp. 105–123.

Lahsen, M. (2005) "Seductive simulations? uncertainty distribution around climate models," *Social Studies of Science* 35(6), pp. 895–922.

Laplace, P.-S. (1902) *A Philosophical Essay on Probabilities*, New York, NY: Wiley (French original 1812).

Latour, B. (1987) *Science in Action: How to Follow Scientists and Engineers through Society*, Cambridge, MA: Harvard University Press.

Lewis, J. M. (1998) "Clarifying the dynamics of the general circulation, Phillips's 1956 experiment," *Bulletin of the American Meteorological Society* 79, pp. 39–60.

Lövbrand, E., J., Stripple and B. Wiman (2009) "Earth system governmentality," *Global Environmental Change* 19(1), pp. 7–13.

Lövbrand, E., S. Beck, J. Chilvers, T. Forsyth, J. Hedrén, M. Hulme, R. Lidskog and E. Vasileiadou (2015) "Who speaks for the future of Earth? How critical social science can extend the conversation on the Anthropocene," *Global Environmental Change* 32, pp. 211–218.

Lynch, P. (2006), *The Emergence of Numerical Weather Prediction. Richardson's Dream*, Cambridge, MA: Cambridge University Press.

Maas, H. (2014), *Economic Methodology: A Historical Introduction*, New York, NY: Routledge.

Mahony, M. (2014a) "The IPCC and the geographies of credibility," *History of Meteorology* 6, pp. 95–112.

Mahony, M. (2014b) "The predictive state: Science, territory and the future of the Indian climate," *Social Studies of Science* 44(1), pp. 109–133.

Mahony, M. (2015) "Climate change and the geographies of objectivity: The case of the IPCC's burning embers diagram," *Transactions of the Institute of British Geographers* 40, pp. 153–167.

Mahony, M. and M. Hulme (2012) "Model migrations: Mobility and boundary crossings in regional climate prediction," *Transactions of the Institute of British Geographers* 37(2), pp. 197–211.

Mahony, M. and M. Hulme (2016) "Modelling and the nation: Institutionalising climate prediction in the UK, 1988–92," *Minerva: A Review of Science, Learning and Policy* 54, pp. 1–26.

Mathews, A. S. and J. Barnes (2016) "Prognosis: Visions of environmental futures," *Journal of the Royal Anthropological Institute* (N.S.) 22(2), pp. 9–26.

Miller, C. A. (2004) "Climate science and the making of a global political order," S. Jasanoff (ed.) *States of Knowledge: The Co-Production of Science and Social Order*, London, UK: Routledge, pp. 46–66.

Mirowski, P. (2013) *Never Let a Serious Crisis Go to Waste: How Neoliberalism Survived the Financial Meltdown*, London, UK: Verso.

Morgan, M. S. (1990) *The History of Econometric Ideas, Historical Perspectives on Modern Economics*, Cambridge, MA: Cambridge University Press.

Morgan, M. S. (1999) "Learning from models," M. S. Morgan and M. Morrison (eds.) *Models as Mediators: Perspectives on Natural and Social Sciences*, Cambridge, MA: Cambridge University Press.

Nebeker, F. (1995) *Calculating the Weather: Meteorology in the 20th Century*, San Diego, CA: Academic Press.

Novotny, H. (2016) *The Cunning of Uncertainty*, Cambridge, MA: Polity.

Oreskes, N., K. Shrader-Frechette and K. Belitz (1994) "Verification, validation, and confirmation of numerical models in the earth sciences," *Science* 263, pp. 641–646.

Oreskes, N. and E.M. Conway (2010) *Merchants of Doubt: How a Handful of Scientists Obscured the Truth on Issues from Tobacco Smoke to Global Warming*, New York, NY: Bloomsbury.

Oreskes, N. and E. M. Conway (2014) *The Collapse of Western Civilization: A View from the Future*, New York, NY: Columbia University Press.

Petersen, A. C. (2006) *Simulating Nature: A Philosophical Study of Computer-Simulation Uncertainties and Their Role in Climate Science and Policy Advice*, Apeldoorn and Antwerpen, Belgium: Het Spinhuis Publishers.

Phillips, N. (1956) "The general circulation of the atmosphere: A numerical experiment," *Quarterly Journal of the Royal Meteorological Society* 82, pp. 123–164.

Pickering, A. (1995) *The Mangle of Practice: Time, Agency, and Science*, Chicago, IL: Chicago University Press.

Pietruska, J. (2009) *Propheteering: A Cultural History of Prediction in the Guilded Age*, Dissertation, Massachusetts Institute of Technology, Cambridge, MA.

Pietruska, J. (forthcoming) *Looking Forward: Prediction and Uncertainty in Modern America*, Chicago: Chicago University Press.

Porter, T. M. (1995) *Trust in Numbers: The Pursuit of Objectivity in Science and Public Life*, Princeton, NJ: Princeton University Press.

Rescher, N. (1998) *Predicting the Future: An Introduction to the Theory of Forecasting*, New York, NY: SUNY Press.

Richardson, L. F. (2007) *Weather Prediction by Numerical Process*, Cambridge, MA: Cambridge University Press (reprint of the First Edition in 1922).

Rouse, J. (1987). *Knowledge and Power: Toward a Political Philosophy of Science*. Ithaca, NY: Cornell University Press.

Russill, C. (2016) "The climate of communication: From detection to danger," S. O'Lear and S. Dalby (eds.) *Reframing Climate Change: Constructing Ecological Geopolitics*. London: Routledge, pp. 31–51

Sandnes, H. and H. Styve (1992) *Calculated Budgets of Airborne Acidifying Components in Europe, 1985, 1987, 1988, 1989, 1990 and 1991*, (EMEP/MSC-West Report 1/92, Technical Report No. 97), Oslo, Norway: Norwegian Meteorological Institute.

Sarewitz, D. and R. A. Pielke (2000) "Prediction in science and policy," D. Sarewitz, R. A. Pielke and R. Byerly (eds.) *Prediction: Science, Decision Making, and the Future of Nature*, Washington, DC: Island Press, pp. 11–22.

Schmelzer, M. (2016) *The Hegemony of Growth: The OECD and the Making of the Economic Growth Paradigm*, Cambridge, MA: Cambridge University Press.

Schneider, B. and T. Nocke (eds. 2014) *Image Politics of Climate Change: Visualizations, Imaginations, Documentations*, Bielefeld, Germany: Transcript.

Schützenmeister, F. (2009) "Offene großforschung in der atmosphärischen chemie?" J. Halfmann and F. Schützenmeister (eds.) *Organisationen der Forschung. Der Fall der Atmosphärenwissenschaft*, Wiesbaden: Verlag für Sozialwissenschaften, pp. 171–208.

Seefried, E. (2015) *Zukünfte. Aufstieg und Krise der Zukunftsforschung 1945–1980*, Berlin, Germany: De Gruyter.

Shackley, S. (2001) "Epistemic lifestyles in climate change modelling," C. A. Miller and P.N. Edwards (eds.) *Changing the Atmosphere*, Cambridge, MA: MIT Press, pp. 107–134.

Taylor, P. J. and F. H. Buttel (1992) "How do we know we have global environmental problems? Science and the globalization of environmental discourse," *Geoforum* 23(3), pp. 405–416.

Thorpe, A. J., H. Volkerts and M. J. Ziemanski (2002) "The Bjerknes' Circulation Theorem: A historical perspective," *Bulletin of the American Meteorological Society* 84, pp. 471–480.

Tinbergen, J. (1940) "Econometric business cycle research," *Review of Economic Studies* 7, pp. 73–90.

UNECE (United Nation Economic Commission for Europe) (ed. 1999) *Strategies and Policies for Air Pollution Abatement. Major Review Prepared under the Convention on Long-range Transboundary Air Pollution*, New York, NY: United Nations.

Wallerstein, I. et al. (2013) *Does Capitalism Have a Future?* Oxford, UK: Oxford University Press.

Walsh, L. (2013) *Scientists as Prophets: A Rhetorical Genealogy*, Oxford, UK: Oxford University Press.

Part I

Junctions

Science and politics of prediction

3 Calculating the weather

Emerging cultures of prediction in late nineteenth- and early twentieth-century Europe

Gabriele Gramelsberger

Introduction

Numerical weather prediction as we know it today was developed over the course of the twentieth century. It is based on mathematical models for expressing the development of the seven main weather variables: wind velocity in three directions, temperature, pressure, density, and humidity. As unpredictability due to the inaccuracy of measurement data (initial data problem) is an intrinsic part of the weather prediction problem (Lorenz 1963), today's prediction models are restarted every six hours with new measurement data in order to keep the prediction errors small. Thus, weather forecasting has improved over the past years due to the rapid growth of the global infrastructure of weather observation from ground observation to satellite sensing, the enormous computational resources for data analysis, and the permanent computation of weather predictions. This global technological assemblage for predicting weather, which is also used for projecting climate change, is the "vastest machine" mankind has ever installed (Edwards 2010).

However, the main insights for predicting weather changes are rooted in the scientific achievements of the nineteenth century, and these insights are threefold. First, it comprises the knowledge that weather is not a local phenomenon, but caused by air masses traveling around the globe, which are influenced by regional conditions. Therefore, meteorologists had to develop strategies to evaluate local measurements of the main weather observables within a spatial scale of 1,000 to 2,500 km to retrieve relevant information. Only this so-called "synoptic" scale sufficiently describes the emergence of local weather phenomena. Second, the awareness that the atmosphere is a three-dimensional and dynamic medium proved crucial. Thus, "barometry" and "dynametry"—early attempts to theorize and predict weather changes by looking at local changes in air pressure—had to be completed by a global circulation theory of the atmosphere's hydrodynamics. And third, the insight that hydrodynamic theory, neglecting the influence of elements such as humidity, had to be combined with thermodynamic theory. Bringing all three aspects together led to today's numerical weather prediction models.

While the nineteenth century was characterized by the struggle of various approaches to tackle the weather prediction problem, but ultimately paved the way for what is known as "dynamical meteorology," the twentieth century was full of efforts to expand and apply dynamical meteorology to create operational forms of weather and climate prediction services (see also Rosol in this volume; Martin-Nielsen in this volume; Heymann and Hundebøl in this volume). Due to limited computing resources, the actual practice of weather forecasting was based on synoptic methods until the 1970s, although the approach of numerical weather prediction had been explored since the late nineteenth century. Based on this background, the paper will explore the early concepts of weather prediction in Europe with a focus on the contributions of the German-speaking countries, which were very early in recognizing and advancing the dynamical approach.[1] Not only was the synoptic method developed in Germany in the 1820s (Brandes 1820), the shift from descriptive to dynamical meteorology—developed by the American meteorologist, William Ferrel (Ferrel 1856, 1858, 1661, 1877; 1886)—was also supported mainly by German-speaking meteorologists, as William Morris Davis recognized in 1887. "Still, it is only in Germany [and Austria] that [Ferrel's advanced theories of dynamical meteorology] have had much effect on recent text-books, and it is to be feared that even the present work [Ferrel 1886] may not reach the readers who ought to have it" (Davis 1887: 540).

This chapter thus begins with a prehistory of weather prediction in the nineteenth century. It explores the scientific ideas and practices serving weather prediction, which gave rise to different cultures of weather prediction, and the struggle between competing meteorological conceptions in the late nineteenth and early twentieth century (1880–1930). The chapter concludes with a summary of the emerging culture of numerical weather prediction.

The prehistory of modern weather prediction (1800–1880)

Discovering the synoptic scale

In his book, *Meteorology in History*, William Napier Shaw, a leading British meteorologist of the late nineteenth and early twentieth century, stated:

> The invention of the barometer and the thermometer marks the dawn of the real study of the physics of the atmosphere, the quantitative study by which alone we are enabled to form any true conception of its structure.
>
> (Shaw 1919: 115)

These feats were accomplished by Evangelista Torricelli, who had invented the barometer for measuring air pressure in the 1640s, and Daniel Gabriel Fahrenheit, who had improved Galileo Galilei's water thermometer to make the mercury thermometer in the 1710s. As early as 1667, Robert Hooke had outlined an advanced *Method for Making a History of the Weather* for the Royal Society in London based on eight observables. Although Hooke's approach of coordinated weather measurements was progressive in its aspiration to shed light on the underlying "cause and laws of weather" (Hooke 1667: 175), "little came of their efforts and interest in coordinated weather observation declined from about 1660 to the end of the century" (Feldman 1990: 146, 147). It took another century before coordinated weather observations came into increasing practice in Europe and substantial meteorological treaties were published (Hellmann 1883). In 1780, a first international network was introduced by the Societas Meteorologica Palatina, located in Mannheim, which coordinated 39 stations in 14 countries taking measurements in a modern manner using standardized devices at coordinated time periods (Wege and Winkler 2005).

While the amount of measurement data increased substantially during the course of the late eightieth and early nineteenth century, the question of how to evaluate these data to understand weather changes still had to be answered. A first attempt was provided by what was called the "climatological" or "statistical" approach. This approach gained ground when researchers like Alexander von Humboldt and Heinrich Dove analyzed measurement data statistically, connected the daily, monthly and yearly mean values of temperature and air pressure with geographic information, and projected them onto maps (Bezold 1885). As early as 1817, Humboldt had introduced isolines of equal temperature to depict climatological patterns (Humboldt 1817, 1820/1821; Schneider and Nocke 2014), but very soon meteorologists produced incorrect mean values by using too much computation to interpolate the data, as the German meteorologist, Wilhelm von Bezold, reported in an early review on progress in weather forecasting (Bezold 1885). Although the climatological approach gave some insight into the distribution of temperature and precipitation, it failed to derive laws from the various means and their periodicity, in particular for wind and barometric studies. Therefore, another approach was needed, and this approach was called "synoptic." In contrast to the climatological view on weather data, synopsis is a method of "similarity" and of the "moment" (Bezold 1885: 315).

The inventor of the synoptic method was the German professor of mathematics at the University of Breslau, Heinrich W. Brandes. In 1820, he analyzed the international data sets of the Societas Meteorologica Palatina for a storm in 1783 to design a weather map with isobars on the synoptic scale (Shaw 1919: 298 et seq.). Brandes made high pressure areas (called "anticyclones") and low pressure areas (called "cyclones" or "depressions," respectively) visible for the first time. Based on his map he realized that differences in air pressure could be huge within small areas and that these local differences cause wind. Further, he concluded that air streams rectilinearly from all sides towards the barometric minima (Brandes 1820). However, this conclusion was wrong and Brandes' work was heavily criticized by Dove, who was the leading meteorologist of his time in Germany and internationally well known. In his 1837 book on wind theory he disproved Brandes and presented the famous "Dove's rule" that wind in the northern hemisphere rotates clockwise around anticyclones and counterclockwise around cyclones (Dove 1837). Dove's influence discredited Brandes' wind theory, but more importantly, brought the synoptic method into disrepute for nearly forty years.[2] Only in the 1850s was the importance of synoptic studies for weather phenomena rediscovered by the Dutch meteorologist Christoph Buys-Ballot (Buys-Ballot 1854, 1857). Buys-Ballot, who articulated Dove's rule more generally as what he called the "baric wind law," marked the difference between climate and weather; he proved that synoptic maps were the only appropriate method for investigating weather phenomena. For this Buys-Ballot was called the "father of modern meteorology" by meteorologists of his time (Bezold 1885: 319).

Early attempts at weather forecasting

Buys-Ballot's studies in the 1850s took place at a time when two major developments were changing meteorology. The first was the introduction of autographic measurement devices, which continually registered the main meteorological observables and automatically recorded them on paper strips (Jelinek 1850; Middleton 1969). The second development was the introduction of telegraphic weather reports. From January 1858, the French astronomer, Urbain Leverrier, gathered telegraphically communicated weather data from various European countries and released daily weather maps in Paris. Within a decade, meteorological offices were founded in Europe, which established observation networks, exchanged telegraphic weather data, and published regular weather maps.[3] Thanks to autographic devices and telegraphy, an increasingly dense picture of the current weather situation for the lower atmosphere could be gathered, raising hopes that synoptic maps could be used to project the trajectories of the cyclones and anticyclones roaming over the maps of the future.

The major driver for establishing observation networks and developing forecasting methods in the mid-nineteenth century was the need for reliable storm warnings. It was the November 1854 storm on the Baltic Sea, which destroyed nearly the entire French and British fleets, which led Leverrier to establish the telegraphic weather service. Another storm, the Royal Charter storm in 1859, which caused the loss of over 800 lives and the steam clipper Royal Charter, inspired Admiral to the British Navy, Robert Fitzroy, to develop a forecasting method he called "dynametry." He combined local measurement data with statistical and mathematical methods. Fitzroy took into account the complex wind patterns of the mid-latitudes, the northeast and southwest motion that Dove called the "wind poles," assimilating all intermediate directions to the characteristics of these extremes. He traced them back to the polar and tropical currents and distinguished them from local effects. He also considered dynamic forces caused by heat or cold, by the expansion of air masses, or other causes. Furthermore, he was aware that changes in weather and wind were preceded and accompanied by alterations in the state of the atmosphere, and that these alterations were indicated sooner in some places than others. Therefore, changes in temperature, pressure, and wind direction could be seen as "signs of changes [of weather] likely to occur soon" (Fitzroy 1863: 177). On this basis of knowledge and measurement data—compiled from 30 to 40 weather telegrams daily—in the early 1860s, he introduced his concept of weather forecasting to the newly founded Meteorological Department of the Board of Trade, the forerunner of the British Meteorological Office. Fitzroy coined the term "weather forecast," defining it as "strictly applicable to such an opinion as is the result of a scientific combination and calculation" (Fitzroy 1863: 171; Anderson 2005).

Although Fitzroy was sure that the dynametry of air would become a subject for mathematical analysis, accurate formulas, and numerical computation, he had to improve the more accessible tool of weather maps because of the limited capacity of computation in his day. He introduced maps with movable wind markers and "nodes"—central areas around which the principal currents circulate or turn. He used color gradients to mark the energy differences in polar (blue) and tropical (red) air streams. These maps and the knowledge of dynamical principles guided meteorologists in their forecasting work. But this work was more an art than an exact science. It was based mainly on experience and a feel for the dynamical principles than on computation or geometrical construction. Tor Bergeron, a leading Swedish meteorologist of the twentieth century, evaluated synoptic meteorology of the 1860s as "a new-born science and by far not an exact one … [It] had to get on with mainly empirical, formal and one-sided methods, which were quite un-fit to the extreme complexity of its main problems" (Bergeron 1941: 251).

The lack of complexity led to false prognoses and gave synoptic meteorology a bad reputation until the methods of weather analysis and databases advanced in the 1910s (Scherhag 1948). In particular, the Austrian meteorologists Heinrich von Ficker and Albert Defant improved synoptic weather forecast substantially. Ficker developed a theory of cyclogenesis describing patterns of temperature and currents (Ficker 1910, 1911, 1912) long before the polar front theory of the so-called "Bergen school" emerged (Bjerknes 1917; Bjerknes and Solberg 1922). In 1917 Defant published the first text book of synoptic meteorology and forecasting (Defant 1918).

From barometry to hydro- and thermodynamics

Although the reputation of weather forecasting based on synoptic maps declined in the late nineteenth century, the observation of cyclones and anti-cyclones roaming over the maps had proven key for explaining local weather phenomena (Kutzbach 1979). What was needed was a more elaborate theory of cyclogenesis. Besides the meteorological interpretation of barometric minima and maxima, since the seventeenth-century experimental physicists had investigated air pressure. "Barometry"—based on Boyle's law of the influence of pressure on the volume of a gas (Boyle 1660)—had been applied by the English physicist, Edmund Halley, to investigate the relationship between barometric pressure and altitude above sea level (Halley 1686/1687). However, in Halley's studies the atmosphere was treated as an ideal hydrostatic fluid without thermodynamical aspects. It took another century before the actual conditions of the atmosphere were considered by the French and Swiss mathematicians, Daniel Bernoulli and Johann Heinrich Lambert (Bernoulli 1738; Lambert 1765). Bernoulli's and Lambert's work paved the way from experimental physics to meteorology. Lambert was able to show that air's density is directly proportional to its pressure, and inversely to its heat (Lambert 1765: 92 et seq.). However, as Bernoulli concluded his work, "no consistent relation could be found between pressure and height, at least in the lower atmosphere, where heat and vapor content are variable" (Feldman 1985: 142; Pick 1855).[4] The reason was the following: Barometry was a typical case of applied mathematics trying to infer empirical formulas from observations. This resulted in numerous empirical formulas for specific contexts, but as George G. Stokes claimed in 1845 regarding the situation of fluid dynamics at his time:

> Such formulae, although fulfilling well enough the purposes for which they were constructed, can hardly be considered as affording us any material insight into the laws of nature; nor will they enable us to pass from consideration of the phenomena from which they were derived to that of others of a different class, although depending on the same causes.
>
> (Stokes 1845: 76)

Thus, the inductive approach had to be complemented with a deductive one, and the roots of this deductive approach——later called "dynamical meteorology"——were mechanics and theoretical physics expressible by differential equations. Thus, the local, synoptic scale had to be combined with the global view on atmospheric circulation as well as with the perspective of the mechanical laws known since Newtonian times. In other words, as US meteorologist, Cleveland Abbe, lamented, "The professional meteorologist" had to become "a mechanician, mathematician and physicist" (Abbe quoted in Nebeker 1995: 28).

Primers of modern weather forecasting (1880–1930)

The global view of the circulation of the atmosphere

An early understanding of some of the causes of the global atmospheric circulation emerged in the late seventeenth century. In 1686, Halley had explained that solar radiation differs for low and high latitudes (Halley 1686/1687a). Therefore, heated tropical air is replaced by cooler air from polar regions, thus causing a north–south circulation. This circulation is, as the English physicist, George Hadley, had pointed out in 1735, deflected by the Earth's rotation (Hadley 1735). Because the speed of rotation differs at each point on Earth, as Dove had explained (Dove 1837), the deflection of air masses differs as well, causing a difference in rotational speed between moving air masses and the places to which these masses have moved. These differences gradually change the direction of the air currents, for instance when they come from the North Pole, from north to northeast to east. For a valid theory of global atmospheric circulation, two major problems had to be tackled: First, the question had to be answered as to how the exchange between airstreams of tropical and polar regions is organized three-dimensionally. Second, while the tropical "trade winds" are regular and easy to explain, the wind situation of the mid-latitudes is far more complex. Dove tried to understand the complex wind patterns of mid-latitudes, but was later corrected by Ferrel (Ferrel 1856).

From these considerations followed a three-cell model of global circulation for each hemisphere, with the polar cells, the mid-latitude cells (about 30°N to 60°N and 30°S to 60°S) and what is called the "Hadley cell" today (about 30°N to 30°S latitude). The American oceanographer, Matthew Fontaine Maury, had elaborated on such a cell model in 1855, but his explanation of global circulation was "most arbitrary and unreasonable" (Davis 1887: 539; Maury 1855). In contrast, Ferrel took the Coriolis effect of the Earth's rotation into account and rooted the three-cell circulation model in a sound physical foundation (Ferrel 1856, 1858; Fleming 2000, 2002). He was able to show that Dove's explanation of the north-east circulation did not explain the wind patterns of the mid-latitude cells, but that due to the Earth's rotation the general motion is an eastward one, which requires that the entire circulation of the tropical winds be reversed (Ferrel 1856). However, independent of Ferrel, the English scientist, James Thomson, brother of Lord Kelvin, had also reversed the direction of the tropical winds; he further added "that, in temperate latitudes, there are three currents at different heights" bringing air from and back to the poles (Thomson 1857: 38). Ferrel integrated this addition into his model in (Ferrel 1861).

The reason why Ferrel, rather than Thomson, was called the "standard authority" for dynamical meteorology was that Ferrel's circulation model was rooted in advanced mathematics. Between 1877 and 1886 he fully elaborated his mathematical theory in his reports on *Meteorological Researches* and *Recent Advances in Meteorology* (Ferrel 1877; 1886). However, Ferrel's theory was not widely recognized, because he published it in unknown journals and administrative reports, and because his superior mathematics was too advanced for his time. As Davis stated thirty years after Ferrel's first models in 1856 and 1857:

> Some thirty years ago [1850s], Ferrel made the initial steps towards [the general atmospheric circulation's] rational solution; and, with a single exception, there has been no one else working in this profitable field until a few of the European mathematical meteorologists lately entered it.
>
> (Davis 1887: 539)

Ferrel was more recognized in German-speaking countries in the 1880s due to the German meteorologist, Adolf W. Sprung, who had popularized dynamical meteorology and in particular Ferrel's theory with his well-received *Lehrbuch der Meteorologie* (Sprung 1885).

The German debate on a rational mechanics of the atmosphere

Sprung, who worked at the German Sea Observatory in Hamburg from 1876 and who became director of the instrument division of the Prussian Meteorological Institute in 1886 and director of the Meteorological Observatory Potsdam in 1892, was an outstanding meteorologist of the nineteenth century. On the one hand, he tried to apply advanced mathematics to meteorology; on the other, he had a profound grasp of empirical meteorology and became famous for conceiving new meteorological instruments. In his 1885 textbook of meteorology, he conceived a modern image of meteorology based on physical laws, mathematics, and an advanced conceptual model of global circulation. Sprung was aware that the atmosphere was a three-dimensional and dynamic medium, and that observation at his time covered only the situation of the lower atmosphere on the ground. Aerology of what was called the "free atmosphere" had just begun in the 1880s, with kites and balloons supported by the field activities of the *First International Polar Year 1882/1883* (Assmann and Berson 1899, 1900; Höhler 2001). Sprung referred extensively to Ferrel in his chapter on atmospheric circulation and used Ferrel's equations to compute wind velocity for his model of general circulation (Sprung 1885: 205 et seq.). Moreover, he was the intellectual link between the typical meteorologist, who at this time was "only an observer, a statistician, an empiricist" (Abbe reported in Nebeker 1995: 28) and theoretical physicists, who were well trained in mathematics and became increasingly interested in atmospheric circulation. This special situation initiated a debate in Germany in the 1880s between meteorology and physics, documenting the clash between the two scientific cultures just before meteorological theory began to turn into the physics of the atmosphere.

This debate was initiated by the physicist and engineer, Werner von Siemens. Inspired by his brother's lecture "On the Conservation of Solar Energy," read before the Royal Society in London (William Siemens 1881), Werner von Siemens became interested in the energy system of the Earth's atmosphere. In a paper "On the Conservation of Energy in the Earth's Atmosphere," read before the Academy of Science in Berlin in 1885, he posed the question as to why meteorologists do not search for the "first principles" of weather phenomena, but instead talk about local barometric minima and maxima as the sources of wind (Siemens 1886). Of course, Siemens was convinced that only the general approach of theoretical physics would be able to unveil these first principles and that hydro- and thermodynamics would describe atmospheric circulation in all locations. In this paper, he started to develop an ideal model of a stationary atmosphere and successively integrated various causes of circulation, including free and latent heat, the Coriolis effect of the Earth's rotation (without knowing the work of Ferrel), and vertical as well as horizontal differences in the movement of the air due to its density. He believed that meteorologists overestimated the influence of the friction between air and ground for the lower atmosphere, emphasizing instead the influence of the kinetic energy of the free atmosphere. Finally, he concluded that the irregularities of all these intermingled causes would "hamper meaningful weather forecasts for all time" (Siemens 1886: 279; translated by the author)—a statement which was confirmed in 1963 by Edward Lorenz's studies on "Deterministic nonperiodic flow," concluding the unpredictability of weather (Lorenz 1963; see also Heymann and Hundebøl in this volume).

Siemens was not aware of Sprung's and Ferrel's work. Sprung's book was published while Siemens was preparing his paper. Siemens's 1885 paper caused a debate in the *Meteorologische Zeitschrift*—the leading journal of the German-speaking meteorological community. In a comment, Wladimir Köppen, one of the editors, ambiguously lauded Siemens's paper as important support by such an authority for achieving an understanding of "the complex machine called atmosphere," hinting in what was already known to be the right direction (Köppen 1886: 234). In fact, the critique by the meteorologist was extensive. Particularly questioned was Siemens's preference for kinetic energy over the barometric gradient and friction as the main causes of wind. A paper by the meteorologist Max Moeller that appeared shortly thereafter directly attacked Siemens's idea of energy conservation of the atmosphere, emphasizing the loss of energy due to friction between air and the ground (Moeller 1887). In a subsequent paper, Siemens defended his idea by asking Moeller about the scientific foundation of his propositions.

> The author [Moeller] doesn't touch the main question: Whence does the enormous energy of the barometric minima and maxima result? In my opinion the huge source from which the energy—causing the barometric minima and maxima and transforming the partly stagnant air movements close to the ground into stormy movements—results from the permanent circulation of the atmosphere [...]. That friction between air and ground would have such a significant influence on the global circulation—more significant than the conservation of energy—has to be questioned.
> (Siemens 1887: 425; translated by the author)

Siemens's perspective clearly addressed global conditions as causes of local phenomena, while meteorologists addressed the local and synoptical view. His view was supported by a paper by Hermann von Helmholtz in 1888, who used Euler's equation to derive integrals for an incompressible atmosphere neglecting friction. Referring to an older paper he had written on "Discontinuous Movements of Fluids" (Helmholtz 1868), Helmholtz conceived the atmosphere as a set of air layers with different rotational velocities, at the boundaries of which wave effects are caused—sometimes becoming indirectly visible through (stratiform) clouds (Helmholtz 1868, 1888, 1889, 1890). Strangely, Sprung, although referring positively to Helmholtz, criticized Siemens but also Ferrel in an 1890 paper by questioning whether a theory of the general system of winds could be based on theoretical derivations (Sprung 1890). Sprung questioned Siemens's model based on an initial state of the atmosphere with a constant rotational velocity of the air. This, as Sprung said, was the main misinterpretation from Hadley to Dove, and by Siemens as well. Siemens rejected Sprung's attack as a misinterpretation of his ideas and again confirmed his model view and his basic assumption that barometric minima and maxima have no local, but global sources (Siemens 1891).

Besides Siemens and Helmholtz, physicists like Heinrich Hertz (1884) and Anton Oberbeck (1888) contributed substantially to the mechanical and physical understanding of the complex machine called atmosphere. In 1884, Hertz suggested a thermodynamical model and a "Graphical Method for Deriving the Adiabatic Changes of State of Humid Air" (Hertz 1884). In 1888 Oberbeck— inspired by Siemens' theory, which he called a first "rational mechanics of the atmosphere" (Oberbeck 1888: 383; translated by the author)—"for the first time, integrated the equations of motion for fluids under conditions approximating those of the atmosphere" (Abbe 1890: 9). Although Oberbeck dealt with the assumption of an incompressible atmosphere for the equation of continuity like Helmholtz, he corrected this idealization with variations in velocity. He based his purely hydrodynamic model on rectangular coordinates for each point considering velocity in three directions, temperature, pressure, density, friction, and gravitation. His integrals supported Siemens's results for an atmosphere with and without rotation as well as his conclusion "that the circulation of the higher atmosphere is the main source of the wind system" (Oberbeck 1888: 395; translated by the author).

Without going into further detail, the debate illustrated the insurmountable differences between meteorologists applying observations at a synoptic view and physicists used to working with mathematically expressed (rational) mechanical theories that address the earth system as a global and closed model—closed in the sense that energy is conserved within the atmosphere because incoming and outgoing solar radiation are equated. In retrospect, Heinz Fortak, who held the first chair of theoretical meteorology in Germany from 1962 until 1993, evaluated the contributions of physicists at that time "as too advanced" (Fortak 2001: 358). However, deriving a theory based on mathematical inferences required simplifications—usually called "models" like Helmholtz's model of an incompressible atmosphere. "Model thinking" was typical for physicists, but not for late nineteenth-century meteorologists. Nevertheless, model thinking was to become the core for rational methods of weather forecasting and later climate projections.

The Austrian school of dynamical meteorology

Besides the Royal Prussian Institute of Meteorology in Berlin headed by Bezold (1885–1907), the Zentralanstalt für Meteorologie und Geodynamik in Vienna, headed by Julius von Hann (1877–1897), became a leading meteorological institute (Hammerl 2001; Coen 2006). Both institutes had been engaged in the foundation of the Austrian Meteorological Society in 1865 and the German Meteorological Society in 1883 as well as in the merging of the journals of both meteorological societies in 1886 into the *Meteorologische Zeitschrift*, edited by

Hann und the German climatologist Wladimir Köppen. While Bezold wrote about the "Thermodynamics of the Atmosphere," referring to Hertz (1885) and Émile Clapeyron (1834), and the *Meteorology as Physics of the Atmosphere* (Bezold 1888, 1892), Hann's reputation resulted from his work on the foehn wind of the Alps (Hann 1866, 1896). He rooted his foehn theory in a general wind theory based on physics and mathematics, referring to US meteorologist, James Pollard Espy (1841), and the German mathematician, Karl Theodor Reye (1872),—both of whom provided a rational mechanics of storms. Bezold as well as Hann had profound expertise in mathematics and numerical methods.

Between the 1880s and 1920s, two extraordinary Austrian meteorologists worked at the Zentralanstalt: Max Margules (1882–1906) and Felix Maria Exner (1901–1930).[5] Margules was interested in the atmosphere's resonance—also called the "atmosphere's tide"—resulting in barometric fluctuations (1890). He asked whether these fluctuations were periodic, and whether they could be computed based on a mathematical model for a rotating spheroid enclosed in an envelope of air. He used Pierre-Simon de Laplace's scheme for calculating tides and Buys-Ballot's baric law (Margules 1892–1893; Laplace 1775, 1776). Because the barometric fluctuations in the mid-latitudes were too aperiodic to be computed globally on the basis of such a resonance theory, Margules restricted his view to a single quantum of air moving through an adiabatic field of pressure. In 1901, he calculated the work needed to change the state of a quantum of air from motion into equilibrium (Margules 1901). From these considerations resulted his most known works on the energy and the theory of storms (Margules 1903, 1906, 2006). Margules could prove against the then established theory that the energy potential of the horizontal pressure gradient was not sufficient to explain the horizontal wind velocities. In fact, the source of the energy of winds results from the potential and the inner energy (together called by Margules the "total potential energy") of air masses with different temperature. The layering of warm air over cold air then makes the total potential energy available (called "available kinetic energy" by Margules). Margules's model considered not only the three dimensions of the atmosphere and the vertical movement of air, he delivered an advanced view, which was successfully adapted by Lorenz fifty years later—called "available potential energy"—to explain the conservation of global circulation (Lorenz 1955; Pichler 2001). In 1904 Margules was also interested in developing a mathematical method for forecasting weather, but soon became skeptical about the predictability of weather because of the lack of accurate measurement data (Margules 1906). Nevertheless, Aleksandr K. Khrgian, in his *Meteorology. A Historical Survey*, asserted that—similar to Ferrel's theory—Margules's

> ideas, which were expressed in difficult, mathematical (and somewhat ponderous German), did not 'reach' a wide scientific audience. Only in Ekholm (1907) do we find traces of his thoughts. They became universally recognized only much later, in the 1920's and 1930's.
>
> (Khrgian, 1970: 195)

While Margules was working on his theory of storms, Exner made the very first attempt to numerically predict the synoptic change of pressure for one day (Exner 1902, 1906, 1907, 1908). It took him and his colleague Albert Defant several weeks of calculations, although Exner's computational model did not use the so-called "primitive equations" of infinitesimal physics, which Lewis F. Richardson later applied for his *Weather Prediction by Numerical Process* (1922), establishing the twentieth century style of numerical weather prediction by simulation (Fortak 2001: 364). Instead, his model

> was based upon incorporating the hydrostatic and geostrophic approxima-
> tions into the thermodynamic equation. ... Thus, he arrived at] analytically
> tractable approximations of the governing equations to calculate the surface
> pressure change on a regular latitude-longitude grid ... Technically Exner's
> final refined approach amounted to computing manually the advective rate
> of change of a layer of uniform potential temperature.
>
> (Volkert 2007: 425)

However, Exner's prognoses were more accurate than Richardson's.[6] In this style of thinking related to late nineteenth-century rational mechanics, Exner published his text book on *Dynamische Meteorologie* in 1917, which referred to Ferrel, Overbeck, and Helmholtz (Exner 1917, 1925; Exner and Trabert 1912; Davis 2001). In a review of Exner's book, Edgar W. Woolard, an American meteorologist, summed up the situation of the late 1910s. He stated that due to the irregularity of weather phenomena

> we still are forced to rely extensively upon empirical and statistical meth-
> ods; but such methods are of limited power, and it is imperative that every
> possible effort be made to advance further the exact mathematical and
> physical theory of atmospheric processes ... this need is largely met by Felix
> M. Exner's *Dynamische Meteorologie* ... The book forms a most excellent
> summary of our present knowledge of theoretical meteorology.
>
> (Woolard 1927: 19)

Margules and Exner both influenced dynamical meteorology deeply. Although advanced for their times, they were rooted in the "rational mechanics approach" of dynamical meteorology, in contrast to Richardson's "numerical simulation style." The reason is mainly a practical one. The simulation style of primitive equations models requires large-scale computing resources, which were not on hand in the early twentieth century. Without automatic computation—available only from the mid-1940s on—calculations could be carried out on mechanical devices or by hand as was done by Richardson. In the 1920s, Hollerith machines were used for weather data analysis and prediction. In the 1930s, "the weather service of every major European country was analyzing data by means of punched cards ... Tabulating equipment thus made it possible to use many more data than was possible before" (Nebeker 1995: 93).

Another strategy was the application of geometrical and graphical computing methods as Abbe suggested: "The application of numerical tables and computations ... can be advantageously replaced by graphic processes ... I have, therefore, sought for planimetric or equivalent methods" (Abbe 1890: 9).

Later the Norwegian meteorologist Vilhelm Bjerknes tried to advance graphical methods geometrically (Bjerknes 1900) and algebraically (Bjerknes and Sandstorm 1910/1911). He highlighted the importance of a graphical algebra as following:

> The development of proper graphical methods for performing these operations directly upon the charts will be of the same importance for the progress of dynamic meteorology and hydrography as the methods of graphical statistics and of graphical dynamics have been for the progress of technical sciences.
>
> (Bjerknes and Sandström 1910/1911: 69)

But graphical methods required major idealizations. Therefore, the most common strategy was to simplify algebraic models by linearization, based on either linear perturbation theory (Bjerknes 1916) or on geostrophic approximations that assumed an exact balance between the Coriolis effect and the pressure gradient force, as Helmholtz, Siemens, Oberbeck, Margules and Exner did. The advantage of these idealizations was (and still is) that such analytically tractable approximations allow exact algebraic solutions to be found. Although the price was simplification, the benefit was a mathematically supported theoretical understanding of mechanical and physical relations through these models.

Conclusion: An emerging culture of numerical weather prediction

The old dream of weather prediction has been approached in every century since Hooke outlined his advanced "Method for Making a History of the Weather" based on eight observables, "whereby the Cause and Laws of Weather may be found out" (Hooke 1667: 175). Meteorologists have hoped that, as soon as the cause and laws became known, prediction would be possible as Laplace's daemon has promised—aptly addressed by Bjerknes in (1904):

> One will agree that the necessary and sufficient conditions for a rational solution of the problem of meteorological prediction are the following: 1. One has to know with sufficient accuracy the state of the atmosphere at a given time [measurements]. 2. One has to know with sufficient accuracy the laws [hydro- and thermodynamics] according to which one state of the atmosphere develops from another.
>
> (Bjerknes 1904, translated in Bjerknes 2009: 663)

To find these causes and laws, meteorologists first tried to achieve a statistical and climatological understanding of weather phenomena in the seventeenth and eighteenth century. Then they discovered the synoptic scale and created the important epistemic tool of synoptic weather maps in the nineteenth century—still in use until the 1970s as the basis for synoptic weather predictions. Finally, they tried to root weather phenomena in physical laws: in the beginning in the empirical physical laws of barometry and dynametry and later in analytically tractable approximations of rational mechanics, introducing model thinking to meteorology.

One of the major problems with model thinking was (and still is) the conceptual and computational treatment of vorticity. Helmholtz in 1858 and William Thomson (Lord Kelvin) in 1867 had already arrived at an ideal hydrodynamic model for a homogeneous inviscid fluid based on the hypothesis that vorticity depends solely on constant density. In such a fluid, no vorticity can occur unless it already exists (Helmholtz 1858). If vorticity exists, it cannot disappear (Thomson 1867). These results, obviously not empirically confirmable, arose from purely theoretical considerations by both physicists, but did not intersect with meteorology. However, the ideal hypothesis made both models analytically solvable (integrable). But for a model to be in accordance with meteorological needs it must take into account a realistic version of vorticity as well as a compressible atmosphere. In other words: The vorticity of the fluid atmosphere does not depend solely on (constant) density as in Helmholtz's and Thomson's models.

Bjerknes realized in the late 1890s that density in a heterogeneous fluid without any restrictions on compressibility was also dependent on other variables such as temperature and pressure. This "contradicted the well-established theorems of Helmholtz and Lord Kelvin which claimed vortex motions and circulations in frictionless, incompressible fluids are conserved" (Friedman 1989: 19) and established credit for Bjerknes's "circulation theorem" as the beginning of modern meteorology—as it is referred to in most of the literature (e.g. Friedman 1989; Gramelsberger 2009).

However, Bjerknes was not the first who tried to conceive a more realistic model. The German physicist, J. R. Schütz (Schütz 1895a, 1895b), as well as the Polish physicist, Ludwik Silberstein (Silberstein 1896), "extended Helmholtz's vorticity equations to the case of a compressible fluid," and in particular "Silberstein's paper discovered (first) all the fundamental aspects to be discussed by Bjerknes in his famous [1898] paper two years later" (Thrope et al. 2003: 472, 473). In the 1930s, the Swedish meteorologist Carl-Gustaf Rossby and the German meteorologist Hans Ertel rearticulated the vorticity equation, laying the foundation for today's weather models. In 1939, Rossby articulated the conservation of the vertical components of absolute vorticity in currents—a linear model that accounted for the perturbations in the upper wersterlies (Rossby 1939; Byers 1960). In 1942, Ertel described the conservation of potential vorticity and stated that "Bjerknes' circulation theorem is a special case of the new hydrodynamical vorticity theorem" (Ertel 1942a: 385; 1942). In 1949 both "derived another vorticity theorem for barotropic fluids, known as the Ertel-Rossby invariant"

(Névir 2004: 485; Fortak 2004; Ertel and Rossby 1949, 149a). Rossby, in particular, became the leading figure for the "simulation style" of dynamical meteorology in the 1940s, introducing numerical weather prediction to the US at the University of Chicago as well as to Europe at the University of Stockholm (Allen 2001; Harper 2008).

Yet the problem with these more realistic models is that they are not analytically solvable due to their intrinsic non-linearities. This was the reason for simplifying and thus idealizing fluid dynamics models, called the "rational mechanics approach" by Oberbeck. With the advent of electronic computers more realistic models could be applied, but early computers like ENIAC were slow and even the very first computational model was just a simple barotropic model with geostrophic wind (Charney et al. 1950). In a barotropic model, pressure is solely a function of density; fields of equal pressure (isobars) run parallel to fields of equal temperature (isotherms), and the geostrophic wind moves parallel to the fields of equal pressure (isobars). These simplifications were necessary to derive an effectively computable model at that time. Furthermore, the numerical simulation of this first computational model was carried out on a space interval of 736 km and a grid consisting of 15 x 18 space intervals for one horizontal layer of the 500 mb contour surface, corresponding to a height of about 5,500 m (Charney et al. 1950). Even for this single level, ENIAC had to carry out more than 200,000 operations. While Neumann later optimistically claimed that these results of the new "simulation style" were as good as the results "subjective" forecasters could achieve (Neumann 1954: 266), others like Norbert Wiener had their doubts, commenting that "500 mb geopotential is not weather" (reported by Arakawa 2000: 6). Nevertheless, the numerical approach to weather prediction quickly became a dominant practice and led to the emergence of the culture of numerical weather prediction in the following years.

This new culture of prediction was based on the belief that weather prediction will be successful "when a complete diagnosis of the state of the atmosphere will be available" and when "there will be no insurmountable mathematical difficulties" of computation and, in particular, no limitations of computing resources (Bjerknes 1904/2009: 663, 666). Today we know that these requirements can be never fully met; what is more problematic is that the intrinsic complexity of not only weather phenomena, but also of the numerically simulated hydro- and thermodynamical models, does not allow for any accurate predictions, no matter how many measurement and computing resources are available. Only Siemens addressed this problem, when he argued that the irregularities of weather phenomena would "hamper meaningful weather forecasts for all time" (Siemens 1886: 279). Lorenz's numerical studies of the chaotic behavior of a weather model confirmed this insight (Lorenz Edward 1963; see also Heymann and Hundebol in this volume). Nevertheless, the simulation practice introduced first by Richardson dominates today's weather predictions and climate projections and forms the core of powerful cultures of numerical weather prediction and of climate prediction.

Notes

1   The history of numerical weather prediction in the twentieth century has been well studied for the United States and partly for the Scandinavian countries, due to their influence on US meteorology in the early twentieth century (e.g. Nebeker 1995; Friedman 1989; Harper 2008; Edwards 2010; Dahan 2010). The other European developments have been a less prominent topic of historical investigations.

2   The German climatologist, Wladimir Köppen, rehabilitated Brandes in a paper in the *Meteorologische Zeitschrift*. He deemed that the influence of Dove had delayed weather studies in Germany by more than forty years (Köppen 1885).

3   In 1847, the Royal Prussian Institute of Meteorology (Königliche Preußische Institut für Meteorologie) was founded as part of the Royal Statistical Bureau in Berlin. In 1851, the Austrian Institute for Meteorology and Geomagnetism (Zentralanstalt für Meteorologie und Erdmagentismus) followed in Vienna and in 1954 the Meteorological Department of the Board of Trade—later called the Meteorological Office—in London. In 1877, the French Meteorological Office (Office Météorologique National de France) in Paris was established. In August 1872, a first international meteorological meeting with fifty scientists from eight states in Europe and the US was held in Leipzig, which became the official conference of directors of meteorological offices. A year later, in September 1873, the first International Conference of meteorology took place in Vienna, establishing the International Meteorological Organization (IMO), the forerunner of today's World Meteorological Organization (WMO) (Steinhauser 2000).

4   Important studies by French physicists like Horace Bénédict de Saussure's *Essais sur l'hygrométrie* and Jean André Deluc's *Idées sur la météorologie* enhanced barometry substantially and paved the way to the nineteenth century mechanical theory of heat as part of meteorology (Saussure 1783; Deluc 1786).

5   Max Margules was a student of Wilhelm von Bezold in Berlin (1879–1880) before he became an assistant at the Zentralanstalt in Vienna from 1882 until 1906 under the directorship of Julius von Hann. Felix Maria Exner was an assistant at the Zentralanstalt in Vienna (1901–1916) under the directorships of Josef Maria Pernter (1897–1908) and Wilhelm Trabert (1909–1915); after Trabert's retirement he became director of the Zentralanstalt from 1916 until 1930.

6   Felix Exner's calculations are considered as "a first attempt at systematic, scientific weather forecasting" (Shields and Lynch 1995: 3). Exner himself in 1908 stated—similar to Vilhelm Bjerknes in 1913—that despite imperfect prognosis it is a major achievement for a weather forecaster to operate on a sound basis that allows him to render his ideas more precisely (Exner 1908, 1995; Bjerknes 1913).

Acknowledgements

I would like to thank the meteorologists Heinz Fortak (Berlin) and Helmut Pichler (Innsbruck) for their personal communications on the history of meteorology in order to better understand the scientific background of the developments described. Besides their theoretical work, both have contributed substantially to the history of meteorology in Germany and Austria (Fortak, 1984, 1988, 1999, 2001, 2004; Pichler 2001, 2004, 2012).

References

Abbe, C. (1890) *Preparatory Studies for Deductive Methods in Storm and Weather Predictions*, Washington, DC: Government Printing Office.

Allan, D. R. (2001) "The Genesis of Meteorology at the University of Chicago," *Bulletin of the American Meteorological Society* 82(9), pp. 1905–1909.

Anderson, K. (2005) *Predicting the Weather: Victorians and the Science of Meteorology*, Chicago, IL: The University of Chicago Press.

Arakawa, A. (2000) "A Personal Perspective on the Early Years of General Circulation Modeling at UCLA," D. A. Randall (ed.) *General Circulation Model Development*, San Diego, CA: Academic Press, pp. 1–65.

Assmann, R. and A. Berson (1899, 1990) *Wissenschaftliche Luftfahrten*, (3 vols.), Braunschweig, Germany: Vieweg.

Bergeron, T. (1941) "A New Era in Teaching Synoptic Meteorology," *Geografiska Annale* 23, pp. 251–275.

Bernoulli, D. (1738) "Hydrodynamics," D. Bernoulli and J. Bernoulli (1968) *Hydrodynamics and Hydraulics*, New York, NY: Dover, pp. 226–229.

Beyers, H. B. (1960) *Carl-Gustav Avrid Rossby, 1898–1957. A Biographical Memoir,* Washington, DC: National Academy of Science.

Bezold, W. von (1985) "Ueber die Fortschritte der wissenschaftlichen Witterungskunde während der letzten Jahrzehnte," *Meteorologische Zeitschrift* 2(9), pp. 313–324.

Bezold, W. von (1888) "Zur Thermodynamik der Atmosphäre," *Königliche Preussische Akademie der Wissenschaften zu Berlin, Sitzungsberichte* 14, pp. 485–522 and pp. 1189–1206.

Bezold, W. von (1892) *Die Meteorologie als Physik der Atmosphäre*, Berlin, Germany: Paetel.

Bjerknes, V. (1898) "Über Einen Hydrodynamischen Fundamentalsatz und Seine Anwendung Besonders auf die Mechanik der Atmosphäre und des Weltmeeres," *Kungliga Svenska Vetenskapsakademiens Handlingar Stockholm* 31, pp. 1–35.

Bjerknes, V. (1904) "Das Problem der Wettervorhersage, Betrachtet von Standpunkt der Mechanik und Physik," *Meteorologische Zeitschrift* 21(1), pp. 1–7 (English translation in Bjerknes 2009).

Bjerknes, V. (1913) "Die Meteorologie als exakte Wissenschaft," (inaugural lecture at the University of Leipzig on 8 January 1913), Braunschweig, Germany: Vieweg.

Bjerknes, V. (1916) "Über Wellenbewegungen in Kompressiblen Schweren Flüssigkeiten," *Abhandlungen der Königlich-Sächsischen Gesellschaft der Wissenschaften, Mathematisch-physikalische Klasse* 35(2), pp. 35–65.

Bjerknes, V. (1917) "Über die Fortbewegung der Konvergenz- und Divergenzlinien," *Meteorologische Zeitschrift* 34(2), pp. 345–349.

Bjerknes, V. (2009) "The Problem of Weather Prediction, Considered from the Viewpoints of Mechanics and Physics (1904)," *Meteorologische Zeitschrift* 18(6), pp. 663–667.

Bjerknes, V. and J. W. Sandström (1910/1911) *Dynamic Meteorology and Hydrography*, (2 vols.), Washington, DC: Carnegie Institution of Washington.

Bjerknes, V. and H. Solberg (1922) "Life Cycle of Cyclones and Polar front Theory of Atmospheric Circulation," *Geophysisks Publikationer* 3(1), pp. 3–18.

Boyle, R. (1660): *New Experiments Physico-Mechanicall, Touching the Spring of the Air and its Effects*, Oxford, UK: H. Hall.

Brandes, H. W. (1820) *Untersuchungen über den mittleren Gang der Wärmeänderungen durchs ganze Jahr; über gleichzeitige Witterungs – Ereignisse in weit voneinander entfernten Weltgegenden; über die Formen der Wolken, die Entstehung des Regens und der Stürme; und über andere Gegenstände der Witterungskunde*, Leipzig, Germany: Barth.

Buys-Ballot, C. (1854) "Erläuterung einer graphischen Methode zur gleichzeitigen Darstellung der Witterungserscheinungen an vielen Orten, und Aufforderung der Beobachter das Sammeln der Beobachtungen an vielen Orten zu erleichtern," *Poggendorfs Annalen* 4, pp. 559–576.

Buys-Ballot, C. (1857) "Note sur le Rapport de l'Intensité et de la Direction du Vent avec les Écarts simultanés du Baromètre," *Comptes Rendus Hebdomadaires*, pp. 765–768.

Charney, J. G., J. von Neumann and R. Fjørtoft (1950) "Numerical Integration of the Barotropic Vorticity Equation," *Tellus* 2, pp. 237–254.

Clapeyron, É. (1834) "Mémoire sur la Puissance Motrice de la Chaleur," *Journal de l'Ecole Royale Polytechnique*, pp. 153–191.

Coen, D. R. (2006) "Scaling Down: The 'Austrian' Climate between Empire and Republic," J. R. Fleming, V. Jankovic and D. R. Coen (eds.) *Intimate Universality: Local and Global Themes in the History of Weather and Climate*, Sagamore Beach, MA: Science History Publications, pp. 115–140.

Dahan, A. (2010) "Putting the earth System in a Numerical Box? The Evolution from Climate Modeling Toward Climate Change," *Studies in History and Philosophy of Modern Physics* 41(3), 282–292.

Davis, W. M. (1887) "Book Review: Advances in Meterology," *Science* 9(226), pp. 539–541.

Davies, H. C. (2001) "Vienna Founding of Dynamical Meteorology," C. Hammerl et al. (eds.) *Die Zentralanstalt für Meteorologie und Geodynamik 1851–2001*, Graz: Leykam, pp. 301–312.

Defant, A. (1918) *Wetter und Wettervorhersage*, Leipzig, Germany: Deuticke.

Deluc, J. A. (1786) *Idées sur la Météorologie*, (2 vols.), London: Spilsbury.

Dove, H. W. (1837) *Meteorologische Untersuchungen*, Berlin, Germany: Sander.

Edwards, P. (2010) *A Vast Machine: Computer Models, Climate Data, and the Politics of Global Warming*, Cambridge, MA: MIT Press.

Ertel, H. (1942) "Ein neuer Hydrodynamischer Wirbelsatz," *Meteorologische Zeitschrift* 59(2), pp. 277–281.

Ertel, H. (1942a) "Über das Verhältnis des neuen Hydrodynamischen Wirbelsatzes zum Zirkulationstheorem von V. Bjerknes," *Meteorologische Zeitschrift* 59(3), pp. 385–387.

Ertel, H. and C.-G.Rossby (1949) "Ein neuer Erhaltungssatz der Hydrodynamik," *Sitzungsberichte der deutschen Akademie der Wissenschaften zu Berlin, Mathematisch-Naturwissenschaftliche Klasse* 1, pp. 3–11 (translated in Ertel and Rossby 1949a).

Ertel, H. and C.-G.Rossby (1949a) "A New Conservation-Theorem of Hydrodynamics," *Geofisica Pura e Applicata* 16, pp. 189–195.

Espy, J. P. (1841) *The Philosophy of Storms*, Boston, MA: Little and Brown.

Exner, F. M. (1902) "Versuch einer Berechnung der Luftdruckänderung von einem Tage zum nächsten," *Sitzungsberichte der Akademie der Wissenschaften zu Wien* 111, pp. 704–725 (translated and introduced by Shields and Lynch 1995).

Exner, F. M. (1906) "Grundzüge einer Theorie der synoptischen Luftdruckveränderung. 1. Mitteilung," *Sitzungsberichte der Akademie der Wissenschaften zu Wien* 115, pp. 1171–1246.

Exner, F. M. (1907) "Grundzüge einer Theorie der synoptischen Luftdruckveränderung. 2. Mitteilung," *Sitzungsberichte der Akademie der Wissenschaften zu Wien* 116, pp. 1819–1854.

Exner, F. M. (1908) "Über eine erste Annäherung zur Vorausberechnung synoptischer Wetterkarten," *Meteorologische Zeitschrift* 25(2), pp. 57–67.

Exner, F. M. (1917) *Dynamische Meteorologie*, Leipzig, Germany: Teubner.

Exner, F. M. (1925) *Dynamische Meteorologie*, (extended and improved reprint), Wien, Austria: Springer.

Exner, F. M. and W. Trabert (1912) "Dynamische Meteorologie," *Encyklopädie der mathematischen Wissenschaften* VI-1B(3), pp. 179–234.

Feldman, T. S. (1985) "Applied Mathematics and the Quantification of Experimental Physics: The Example of the Barometric Hypsometry," *Historical Studies in the Physical Sciences* 15, pp. 127–197.

Feldman, T. S. (1990) "Late Enlightenment Meteorology," T. Frängsmyr, J.L. Heilbron and R.E. Rider (eds.) *The Quantifying Spirit in the 18th Century*, Berkeley, CA: University of California Press, pp. 143–179.

Ficker, H. (1910) "Die Ausbreitung kalter Luft in Rußland und Nordasien," *Sitzungsberichte der Akademie der Wissenschaften zu Wien* 119, pp.1769–1837.

Ficker, H. (1911) "Das Fortschreiten der Erwärmung (Wärmewellen) in Rußland und Nordasien," *Sitzungsberichte der Akademie der Wissenschaften zu Wien* 120, pp. 745–836.

Ficker, H. (1912) "Kälte- und Wärmewellen in Nordrußland und Asien," *Meteorologische Zeitschrift* 29(2), pp. 378–383.

Fleming, J. R. (2000) *Meteorology in America, 1800–1870*, Baltimore, MD: Johns Hopkins University Press.

Fleming, J. R. (2002) "History of Meteorology," B.S. Biagre (ed.) *A History of Modern Science and Mathematics*, (vol. 3), New York, NY: Scribner's, pp. 184–217.

Ferrel, W. (1856) "An Essay on the Winds and the Currents of the Ocean," *Nashville Journal of Medicine and Surgery* 11, pp. 287–301.

Ferrel, W. (1858) "The Influence of the Earth's Rotation upon the Relative Motion of Bodies near its Surface," *Astronomical Journal* 5(109), pp. 97–100.

Ferrel, W. (1861) "The Motion of Fluids and Solids on the Earth's surface," *American Journal of Science*, Series 2, 31(91), pp. 27–50.

Ferrel, W. (1877) *Meteorological Researches for the use of the Coast Pilot*, Washington, DC: US Coast Service.

Ferrel, W. (1886) *Recent Advances in Meteorology*, (Annual Report of the Chief Signal Officer), Washington, DC: US War Department.

Fitzroy, R. (1863) *The Weather Book: A Manual of Practical Meteorology*, London: Longman and Green.

Fortak, H. and P. Schlaack (1984) *100 Jahre Deutsche Meteorologische Gesellschaft in Berlin*, Berlin, Germany: Deutsche Meteorologische Gesellschaft.

Fortak, H. (1988) *Gedächtniskolloquium für K.H. Hinkelmann am 14. Mai 1987 in Mainz*, (Annalen der Meteorologie 24), Offenbach: Deutscher Wetterdienst.

Fortak, H. (1999) *50 Jahre Institut für Meteorologie der Freien Universität Berlin*, Berlin, Germany: Berliner Wetterkarte e.V.

Fortak, H. (2001) "Felix Maria Exner und die österreichische Schule der Meteorologie," C. Hammerl et al. (eds.) *Die Zentralanstalt für Meteorologie und Geodynamik 1851–2001*, Graz, Austrai: Leykam, pp. 354–386.

Fortak, H. (2004) "Hans Ertel's Life and his Scientific Work," *Meteorologische Zeitschrift* 13(6), pp. 453–464.

Friedman, M. (1989) *Appropriating the Weather. Vilhelm Bjerknes and the Construction of a Modern Meteorology*, Ithaca, NY: Cornell University Press.

Gramelsberger, G. (2009) "Conceiving Meteorology as the Exact Science of the Atmosphere – Vilhelm Bjerknes' Revolutionary Paper of 1904," *Meteorologische Zeitschrift* 18(6), pp. 663–667.

Hadley, G. (1735): "The Cause of the General Trade-Wind," *Philosophical Transactions of the Royal Society of London* 29, pp. 58–62.

Hann, J. von (1866) "Zur Frage über den Ursprung des Foehn," *Zeitschrift der Oesterreichischen Gesellschaft für Meteorologie* 1(1), pp. 257–263.

Hann, J. von (1896): *Die Erde als Ganzes, ihre Atmosphäre und Hydrosphäre*, Prague, Czech Republic: Tempsky

Halley, E. (1686/1667) "A Discourse of the Rule of the Decrease of the Height of the Mercury in the Barometer," *Philosophical Transactions of the Royal Society of London* 16, pp. 104–116.

Halley, E. (1686/1667a) "An Historical Account of the Trade Winds, and Monsoons, Observable in the Seas between and Near the Tropicks, with an Attempt to Assign the Phisical Cause of the Said Wind," *Philosophical Transactions of the Royal Society of London* 16, pp. 153–168.

Hammerl, C., W. Lenhardt, R. Steinacker and P. Steinhauser (eds. 2001) *Die Zentralanstalt für Meteorologie und Geodynamik 1851–2001. 150 Jahre Meteorologie und Geophysik in Österreich, 1851–2001*, Graz, Austria: Leykam.

Harper, K. C. (2008) *Weather by the Numbers. The Genesis of Modern Meteorology*, Cambridge, MA: MIT Press.

Hellmann, G. (1883) *Repertorium der Deutschen Meteorologie. Leistungen der Deutschen in Schriften, Erfindungen und Beobachtungen auf dem Gebiete der Meteorologie und des Erdmagnetismus von den ältesten Zeiten bis zum Schlusse des Jahres 1881*, Leipzig, Germany: Engelmann.

Helmholtz, H. von (1858) "Ueber Integrale der hydrodynamischen Gleichungen, welche den Wirbelbewegungen entsprechen," *Journal für die reine und angewandte Mathematik* 55, pp. 25–55.

Helmholtz, H. von (1868) "Ueber discontinuirliche Flüssigkeitsbewegungen," *Monatsbericht der Koeniglich Preussischen Akademie der Wissenschaften zu Berlin*, pp. 215–228.

Helmholtz, H. von (1876) "Wirbelstürme und Gewitter," *Deutsche Rundschau* 6, pp. 363–380.

Helmholtz, H. von (1888) "Ueber atmosphärische Bewegungen," *Sitzungsberichte der Königlich Preußischen Akademie der Wissenschaften zu Berlin*, 1. Halbband, pp. 647–663.

Helmholtz, H. von (1889) "Ueber atmosphärische Bewegungen. Zweite Mitteilung. Zur Theorie von Wind und Wellen," *Sitzungsberichte der Königlich Preußischen Akademie der Wissenschaften zu Berlin*, 2. Halbband, pp. 761–780.

Helmholtz, H. von (1890) "Die Energie der Wogen und des Windes," *Sitzungsberichte der Königlich Preußischen Akademie der Wissenschaften zu Berlin*, 2. Halbband, pp. 853–872.

Hertz, H. (1884) "Graphische Methode zur Bestimmung der adiabatischen Zustandsänderungen feuchter Luft," *Meteorologische Zeitschrift* 1(11/12), pp. 412–431.

Höhler, S. (2001) *Luftfahrtforschung und Luftfahrtmythos: Wissenschaftliche Ballonfahrt in Deutschland, 1880–1910*, Frankfurt, Germany: Campus.

Hooke, R. (1667) "A Method for Making the History of the Weather," T. Sprat (ed.) *The History of the Royal Society of London, for the Improving of Natural Knowledge*, London: Martyn, pp. 173–179.

Humboldt, A. von (1817) "Memoire sur les lignes iso-thermes," *Annales de Chimie et de Physique* 5, pp. 102–111.

Humboldt, A. von (1820/1821) "On Isothermal Lines, and the Distribution of Heat over the Globe," *The Edinburgh Philosophical Journal* III–V(5–9), pp. 1–20, pp. 256–274, pp. 23–37, pp. 262–281 and pp. 28–39.

Jelinek, C. (1850) "Beiträge zur Construction selbstregistrierender meteorologischer Apparate," *Sitzungsberichte der Akademie der Wissenschaften Wien*, pp. 3–42.

Khrgian, A. (1970) *Meteorology. A Historical Survey*, Jerusalem, Israel: Israel Program for Scientific Translations.

Köppen, W. (1885) "Die Stellung von H. W. Brandes und H. W. Dove, 1820 und 1868, zum barischen Windgesetz," *Meteorologische Zeitschrift* 2(11), pp. 414–416.

Köppen, W. (1886) "Werner Siemens: Ueber die Erhaltung der Kraft im Luftmeere der Erde," *Meteorologische Zeitschrift* 3(5), pp. 233–234.

Kutzbach, G. (1979) *The Thermal Theory of Cyclones – A History of Meteorological Thought in the Nineteenth Century*, (Historical Monograph Series), Boston, MA: American Meteorological Society.

Lambert, J. H. (1765) "Abhandlung von den Barometerhöhen und ihren Veränderungen," *Abhandlungen der Bayerischen Akademie der Wissenschaften* 2, pp. 75–182.

Laplace, P. S. (1775) "Recherches sur Plusieurs Points de Systeme du Monde," *Mémoires de l'Académie royale des sciences de Paris* 88, 1778, pp. 71–183.

Laplace, P. S. (1776) "Recherches sur Plusieurs Points de Systeme du Monde (suite)," *Mémoires de l'Académie royale des sciences de Paris* 89, pp. 187–280 and pp. 283–310.

Lorenz, E. (1955) "Available Potential Energy and the Maintenance of the General Circulation," *Tellus* 7(2), pp. 157–167.

Lorenz, E. (1963) "Deterministic Nonperiodic Flow," *Journal of the Atmospheric Sciences* 20(2), pp. 130–141.

Margules, M. (1890) "Über die Schwingungen periodisch erwärmter Luft," *Sitzungsberichte der Kaiserlichen Akademie der Wissenschaften zu Wien, mathematisch-naturwissenschaftliche Klasse IIa* 99, pp. 204–277.

Margules, M. (1892–1893) "Luftbewegungen in einer rotirenden Sphäroidschale," *Sitzungsberichte der Kaiserlichen Akademie der Wissenschaften zu Wien, mathematisch-naturwissenschaftliche Klasse IIa* 101, pp. 597–626, 102, 11–56 and pp. 1369–1421.

Margules, M. (1901) "Über den Arbeitswert einer Luftdruckvertheilung und über die Erhaltung der Druckunterschiede," *Denkschrift der Kaiserlichen Akademie der Wissenschaften zu Wien* 73, pp. 329–345.

Margules, M. (1903) "Über die Energie der Stürme," *Jahrbuch der K.K. Zentralanstalt für Meteorologie* 42, pp. 1–26.

Margules, M. (1904) "Über die Beziehung zwischen Barometerschwankungen und die Kontinuitätsgleichung," S. Meyer (ed.) *Ludwig Boltzmann Festschrift*, Leipzig, Germany: Barth, pp. 585–589.

Margules, M. (1906) "Zur Sturmtheorie," *Meteorologische Zeitschrift* 23(2), pp. 481–497.

Margules, M. (2006) "Über die Temperaturschichtung in stationär bewegter und in ruhender Luft," *Meteorologische Zeitschrift* 23(6), pp. 243–254.

Maury, M. F. (1855) *The Physical Geography of the Sea*, New York, NY: Harper and Brothers.

Middleton, K. W. E. (1969) *Invention of the Meteorological Instruments*, Baltimore, MD: John Hopkins Press.

Moeller, M. (1887) "Über Verluste an äusserer Energie bei der Bewegung der Luft," *Meteorologische Zeitschrift* 4(9), pp. 318–324.

Nebeker, F. (1995) *Calculating the Weather: Meteorology in the 20th Century*, San Diego, CA: Academic Press.

Neumann, J. von (1954) "Entwicklung und Ausnutzung neuerer mathematischer Maschinen," J. von Neumann (1963) *Collected Works. Design of Computers, Theory of Automata and Numerical Analysis*, (vol 5.), Oxford: Pergamon Press, pp. 248–268.

Névir, P. (2004) "Ertel's Vorticity Theorems, the Particle Relabelling Symmetry and the Energy-vorticity-theory of Fluid Mechanics," *Meteorologische Zeitschrift* 13, pp. 485–498.

Oberbeck, A. (1888) "Ueber die Bewegungserscheinungen der Atmosphäre," *Sitzungsberichte der Königlich Preussischen Akademie der Wissenschaften zu Berlin* 14, pp. 383–395 and pp. 1129–1138.

Pichler, H. (2001) "Von Margules zu Lorenz," C. Hammerl, W. Lenhardt, R. Steinacker and P. Steinhauser (eds.) *Die Zentralanstalt für Meteorologie und Geodynamik 1851 – 2001*, Graz: Leykam, pp. 387–397.

Pichler, H. (2004) "Das Wirken Hans Ertels in Österreich und die Bedeutung seines Wirbelsatzes in der alpinen Meteorologie," *Sitzungsberichte der Leibniz-Sozietät* 71, pp. 51–61.

Pichler, H. (2012) "Die Neugründung der 'Meteorologischen Zeitschrift' nach der Wende – das Wiederaufleben einer alten Tradition," *Bulletin der Österreichischen Gesellschaft für Meteorologie* 1, pp. 19–23.

Pick, A. J. (1855) "Ueber die Sicherheit barometrischer Höhenmessungen," *Sitzungsberichte der Akademie der Wissenschaften Wien*, pp. 415–447.

Reye, K. T. (1872) *Die Wirbelstürme, Tornados und Wettersäulen in der Erd-Atmosphäre mit Berücksichtigung der Stürme in der Sonnen-Atmosphäre*, Hanover, Germany: Ruempler.

Richardson, L. F. (1922) *Weather Prediction by Numerical Process*, Cambridge, MA: Cambridge University Press.

Rossby, C.-G. (1939) "Relation between Variations in the Intensity of the Zonal Circulation of the Atmosphere and the Displacements of the Semi-Permanent Pressure Systems," *Journal of Marine Research* 2, pp. 38–55.

Saussure, H. B. de (1783): *Essais sur l'hygrométrie*, Neuchâtel: Chez S. Fauche père et fils.

Scherhag, R. (1948) *Neue Methoden der Wetteranalyse und Wetterprognose*, Berlin: Springer.

Shields, L. and P. Lynch (1995) "A First Approach Towards Calculating Synoptic Forecast Charts," *Historical Note* 1, Dublin: Meteorological Service, p. 36.

Schneider, B. and T. Nocke (eds., 2014) *Image Politics of Climate Change. Visualizations, Imaginations, Documentations*, Bielefeld, Germany: Transcript.

Schütz, J. R. (1895a) "Über die Herstellung von Wirbelbewegungen in idealen Flüssigkeiten," *Annalen der Physik* 292(9), pp. 144–147.

Schütz, J. R. (1895b) "Über eine bei der theoretischen Einführung incompressibler Flüssigkeiten gebotene Vorsicht," *Annalen der Physik* 292(9), pp. 148–150.

Shaw, N. (1919) *Manual of Meteorology. Meteorology in History*, London: Cambridge University Press.

Siemens, W. von (1881) "On the Conservation of Solar Energy," *Philosophical Transactions of the Royal Society London* 33, pp. 389–398.

Siemens, W. von (1886) "Ueber die Erhaltung der Kraft im Luftmeere der Erde," *Annalen der Physik* 264(6), pp. 263–281.

Siemens, W. von (1887) "Zur Frage der Luftströmung," *Meteorologische Zeitschrift* 4(12), pp. 425–428.

Siemens, W. von (1890) "Zur Frage der Ursachen der atmosphärischen Ströme," *Meteorologische Zeitschrift* 8(3), pp. 336–337.

Siemens, W. von (1891) "Ueber das allgemeine Windsystem der Erde," *Annalen der Physik* 278(2), pp. 257–268.

Silberstein, L. (1896) "Über die Entstehung von Wirbelbewegungen in einer reibungslosen Flüssigkeit," *Bulletin international de l'Académie des Sciences de Cracovie, Comptes rendus des séances de l'année*, pp. 280–290.

Sprung, A. W. (1885) *Lehrbuch der Meteorologie*, Hamburg, Germany: Hoffman und Campe.

Sprung, A. W, (1890) "Ueber die Theorien des allgemeinen Windsystems der Erde, mit besonderer Rücksicht auf den Antipassat," *Meteorologische Zeitschrift* 7(5), pp. 161–177.

Steinhauser, P. (2000) "50 Jahre meteorologische Weltorganisation – zur Bedeutung und Entwicklung der internationalen Zusammenarbeit in der Meteorologie," *ÖGM Bulletin* 1, pp. 12–21.

Stokes, G. G. (1845) "On the Theories of the Internal Friction of Fluids in Motion," G. G. Stokes (1880): *Mathematical and Physical Papers*, (vol. 1), Cambridge, MA: Cambridge University Press, pp. 75–115.

Thomson, J. (1857) "On the Grand Currents of Atmospheric Circulation," (Report of the British Association Meeting Dublin), pp. 38–39.

Thomson, J. (1892) "Bakerian Lecture: On the Grand Currents of Atmospheric Circulation," *Philosophical Transactions of the Royal Society of London A* 183, pp. 653–684.

Thomson, W. (Lord Kelvin) (1867) "On Vortex Atoms," *Proceedings of the Royal Society of Edinburgh* 6, pp. 94–105.

Thrope, A., H. Volkert and M.J. Ziemianski (2003) "The Bjerknes' Circulation Theorem. A Historical Perspective," *Bulletin of the American Meteorological Society* 84, pp. 471–480.

Volkert, H. (2007) "Felix Maria Exner-Ewarten," N. Koertge (ed.) *New Dictionary of Scientific Biography*, New York, NY: Charles Scribner's Sons, pp. 425–427.

Wege, K. and P. Winkler (2005) "The Societas Meteorologica Palatina (1780–1795) and the Very First Beginnings of Hohenpeissenberg Observatory," S. Emeis and C. Lüdecke (eds.) *From Beaufort to Bjerknes and Beyond: Critical Perspectives on Observing, Analyzing, and Predicting Weather and Climate*, Augsburg: Rauner, pp. 45–54.

Woolward, E.H. (1927) "F. M. Exner on Dynamical Meteorology," *Monthly Weather Review* 55(1), pp. 18–20.

4 Which design for a weather predictor? Speculating on the future of electronic forecasting in post-war America

Christoph Rosol

Introduction

Immediately following the end of World War II and amid the dismantling of a well-financed research infrastructure that contributed crucially to the Allied victory, a debate ensued among US scientists and engineers as to the further development and use of electronics for civilian purposes. Chief among the proposals was the application of the acquired engineering skills and tools for weather forecasting—a subject as crucial for conducting war as it was for thriving in times of peace.

The following paper describes three distinctively different concepts of automated weather prediction as held by their respective proponents Vladimir Zworykin, Vannevar Bush, and John von Neumann, all three of whom had been engaged more or less directly in wartime electronics research. It further integrates the perspective of practicing meteorology at the time by examining the attitude the US Weather Bureau took in the process. While diminutive in its historiographic format, the correspondence between those three (or four) actors over the course of 1945 and 1946 exquisitely highlights the different views on the very essentials of prediction techniques in what was to become the formative period of not only numerical models of atmospheric motion, but also numerical experiments, computer simulation and cultures of prediction at large.

Much has been written already about the history of numerical weather prediction, and indeed the story itself could be regarded as a classic role model for the transformation of scientific cultures in general—a good reason to regard, as Matthias Heymann (2010: 194) does, the atmospheric sciences as a key discipline in the twentieth century. Numerical treatment turned an empirical data handling practice into a computational science: this is perhaps the epistemological essence of that very transformation.[1] Of course, one should rush to note that handling data in digital environments is just another practice itself, yet one that is more integrative and conceptually recursive. The characteristic interaction and epistemic interdependence between "model-laden data" and "data-laden models" is a key notion in Edwards (2010); Guillemot (2010), too, points out the strong interlinks between models and data in their mutual validation process. In any case, it is obvious that digital computation has become the norm, first in the natural sciences and parts of the social sciences, and followed more recently by the rise of the so-called digital humanities.[2] And since most of these computations involve statistical analysis or integration over a large mass of variables, big-data infrastructures have become a general prerequisite for much of today's research.

The enormity of this technologically driven shift makes it worthwhile to revisit the original sources from time to time, interrogating them again to examine their changing historiographic status. Accordingly, this paper seeks to describe the diverging views on the technical means appropriate to accomplish the marriage of meteorology with the computer as held at the eve of numerical modeling. Originally envisioned by physicist turned meteorologist Vilhelm Bjerknes in 1904 and manually tested by Lewis Fry Richardson during World War I the fate of the numerical program remained pretty unclear. Implementing it into actual hardware and software was only one emerging view of several—and certainly one that was unexpected by most practitioners in the field. According to the then predominant rationale, meteorology in principle was an empirical profession, defying any causal or deterministic descriptions, such that the most exact methods imaginable would merely amount to statistical techniques.

In this chapter, I will thus focus on the peculiar situation of 1945–46 in which the future trajectory of technology and techniques of weather prediction was still wide open, leaving the epistemic status of modeling the large-scale dynamics and behavior of the Earth's atmosphere, that is, the "general circulation", in suspense. To some extent the discussion presented in this historical snapshot can be regarded as indicative of everything that was to follow, that is, the first test runs on the predictive machine in 1950 (Platzman 1979; Lynch 2008), spurring further theory building on the general circulation, but also a few years later when numerical weather prediction became operative and the first "numerical experiments" on the general circulation finally turned the computer into a heuristic tool for climatology, redefined as a "long-range" or even "infinite forecast" (Neumann 1960; Lewis 2000).

Wartime research explicitly turned basic science into feasibility studies, advancing an engineering approach that moved it away from representing (the laws of nature) toward intervening (into nature). Most prominent in this regard was the creation of the atomic bomb during the Manhattan Project, but several other technical developments also appear to manifest this move, among them aviation, servomechanic controls, ultrahigh frequency radar, vacuum tube applications for all sorts of signal generation and amplification, and finally the mechanical or, later, electronic computer (Harvard Mark I, Mark II, ENIAC, Colossus, Zuse Z4) as both an instrument and a research object itself (Rosol 2007). In turn, these scattered developments in the design of new technologies and especially media technologies (like electronic control circuits and computers) had decisive impacts on theory as manifested by information theory and cybernetics—by origin, or legend, a predictive science[3]—both of which wove the theoretical fabric that girds much of today's computational sciences. Drawing on this hands-on mindset of the wartime years, and spurred by the practical need for well-established aerology in times of incipient commercial aviation, the most intractable part of nature seemed just the right thing to attack, namely the ephemeral and nonlinear fluid dynamics of the atmosphere, or to put it more mundanely: the flows of air across the globe.

This is not to say that simulation itself became a new appropriation of the world. Quite to the contrary, it is an old practice, having always bridged the impenetrable gap between phenomena and model (Serres 2003). Instead the transformation seemed more technological than epistemic. While physical concepts exhibit an astonishing transhistorical continuity, the technical and technological realms in which these concepts unfold shift decisively.[4] And indeed, there is hardly a historical instance where this shift is better visible than in post-war meteorology.

Zworykin's proposal of 1946

On 18 October 1946, the meteorologist Harry Wexler reported to his supervisor Francis Reichelderfer, director of the US Weather Bureau, on his business trip to Princeton three days earlier. Wexler was head of the Special Scientific Services Division at the Weather Bureau and, as such, liaison of the national weather service to the nascent "Meteorological Computing Project," which had been recently established at the idyllically located Institute for Advanced Study (IAS). At the invitation of project leader and world-renowned mathematician John von Neumann, Wexler had traveled repeatedly from Washington to Princeton over the previous months to inquire about the progress of work at the IAS. The building, which would house the electronic computer and its operating team, Wexler noted, was about one third complete. Furthermore, the computational model of the general circulation was in the making and the required adaptation of the "Richardson equations" was almost ready for trial runs to be executed on the often-idle Electronic Numerical Integrator and Computer, or ENIAC:

> Dr. von Neumann is quite pleased at the progress made by Captain Hunt in tackling the general circulation problem and it is believed that the equations are in such shape that they can be put on the ENIAC in the near future... Since the ENIAC is not being used anywhere near to full capacity, it is believed that this machine can be made available for experimental meteorological runs.
>
> (Wexler 1946)

Gradually, everything seemed to get under way. Wexler further noted:

> The project must be given delicate handling at the beginning but should be brought closer to reality as time progresses. To this end the undersigned has furnished and will continue to furnish advice and summarized meteorological information so that the theoreticians will always have

before them the atmosphere as it actually exists and behaves. Some of the historical material prepared during the war, in particular the upper air data, has already proved to be quite useful in Captain Hunt's project.

(Wexler 1946)

The atmosphere "as it actually exists and behaves" became represented by "historical material," derived from data sets collected under World War II conditions. Provided to the theoreticians, that is, mathematicians, these data sets furnished a readily available reality check for the heuristic attempts to adjust the equations to the machine—a mutual adaptation of theory, hard- and software by which rational prognosis should eventually become possible.

Noteworthy is the last paragraph of Wexler's report, which was, if one follows the annotations of Reichelderfer and Jerome Namias (Wexler's colleague at the Weather Bureau) the one most discussed. Wexler wrote that after visiting the Meteorology Project, he and Air Force Captain Gilbert Hunt, a former professional tennis player and future Markov chain specialist, were invited for dinner at the home of Vladimir Kosma Zworykin, vice president of the Radio Corporation of America (RCA). The famous developer of the iconoscope tube, and, as such, respected as one of the fathers of electronic television, oversaw the design of the memory tubes intended for the IAS computer in the nearby research complex of RCA. Following the recently completed development of the electron microscope, the RCA Princeton Laboratories saw great potential not only in *seeing* electronically with vacuum tubes, but also in *calculating* with them. Yet, the conversation in Zworykin's home was also drawn to a rather macroscopic topic:

> Dr. Zworykin is still interested in weather control, and as a result of some recent visual observations he made off Miami Beach of the build-up of cumulonimbus over what were apparently the Bahamas, he thinks that hurricanes might be prevented from forming by igniting oil on the sea surface in critical areas, thus 'bleeding off' energy by thunderstorms which might otherwise go into hurricane formation. Dr. Zworykin thinks we ought to enlist Navy aid in making more comprehensive measurements of [cumulonimbus] build-ups over islands in the tropical oceans as a preliminary to actual experiments with oil. Arguments both for and against were thoroughly discussed over the vodka.

(Wexler 1946)

The topic Wexler reported on here—and was not to forget for his lifetime (Fleming 2010: 222)—was not new to the Weather Bureau staff. About a year earlier, Zworykin had circulated a short proposal, just under ten pages, in which he laid out his vision for global weather control, "a goal recognized as eventually possible by all foresighted men" (Zworykin 1945b). In this paper, which Wexler had received via Edward U. Condon in January 1946 (Condon 1946),[5] Zworykin fantasized, inter alia, about the extensive use of flamethrowers or even atomic bombs to affect local heat balances and to divert hurricanes or ocean currents. Rainfall could be triggered by shock waves, or the seeding of clouds with ice or dust. Longer-term climatic improvements, in turn, could be achieved by large-scale changes in vegetation and alteration of deserts, mountains and glaciers.

In a very peculiar fashion, this famous pamphlet, which the famous meteorologist Jule Charney used to call "the modest proposal" (Platzman 1987: 54), was the point of departure for a phenomenal historical trajectory that led from immediate post-war America to today's climate modeling and, to some extent, the computational sciences in general. In essence, Zworykin's argument is that of an engineer of signaling equipment. Relatively small amounts of selective energy input might discharge and/or control far greater amounts of energy, thus triggering a desired phenomenon to develop or reverse: "The energy involved in controlling weather would be very much less than that involved in the weather phenomenon itself" (Zworykin 1945b: 4).[6] The analogy here can be seen in the design of the triode valve: Much like the control grid attenuates the electron current, so the controlled detonation of nuclear bombs would attenuate the nascent upflow of water and energy from the sea. Formulated to the extreme, Zworykin's conceptual model arranges the whole tropical Atlantic into a kind of super cathode ray tube, promising an interventionist laboratory to divert all hurricanes between cathode Africa and anode America in a controlled fashion.

However, curing must follow diagnostics. As made clear by Zworykin, prerequisite for any "command and control" communication with the weather would be an exact determination of the aerological situation. Even the mere possibility of any weather modification experiments would be extremely difficult to explore because of the sheer size of the areas and the amounts of energy involved, rendering any field studies practically impossible. "This means that a rapid computing model ... would be indispensable in selecting the areas, type and degree of treatment to be used in the verification of experimental work" (Zworykin 1945b: 6).

Thus, Zworykin's megalomaniac scenarios for weather and climate control presented an essential precondition: a precise and, if possible, globally scaled prediction of the generation and further evolution of weather phenomena by electronic computing devices. Zworykin had concrete ideas about the type and working of these devices:

> Rapid computability makes many experimental manipulations possible in a short time. With the aid of card file selection systems past situations may be examined and experimental judgments facilitated. The combination of rapid computability and rapid statistical reference constitutes a very powerful tool for perfecting the prediction methods, as well as for their application. [...] [S]pecial computing equipment can be built to solve special dynamical equations, perform dynamical extrapolation, and facilitate use of statistical techniques for ready reference to past data on similar situations.

[...] The eventual aim here would be to develop an automatic plotting board to provide a model of the movement and modification of pressure systems, on an accelerated time base.

(Zworykin 1945b: 2–3)

Based on the application of such dynamical, statistical, and graphical methods in weather forecasting, Zworykin explained further, a tiered, global network of data centers should be created to collaboratively predict (and improve) the world's weather. Cast into a global architecture of data, the common goal of weather control might even turn into a catalyst for world peace (Zworykin 1945b: 8).

Two reactions: Bush and von Neumann

Therefore, it seemed only consequent that, before sharing his spectacular vision for future meteorology with actual meteorologists, Zworykin initially approached Vannevar Bush, the "czar" of wartime research that was just about to wind down, sending him a first draft of his paper (Zworykin 1945a). Bush had headed the US Office of Scientific Research and Development (OSRD) and thus was in the capacity to direct almost all military research and development during the war. In his reply, Bush stressed his skepticism as to the true realization of local weather intervention. Nevertheless, he also confessed to a longstanding interest in any kind of mechanized analysis and prediction of weather.[7] An electrical engineer by training and familiar with calculating machines through his own experience and designs, he told Zworykin: "The machines undoubtedly will have to be specially adapted to their purpose" (Bush 1945). His own "Differential Analyzer" seemed rather inappropriate for the needs of dynamic meteorology, as the machine was incapable of handling partial differential equations, let alone nonlinear ones. Another device at his disposal, the "Optical Integraph", appeared more promising in this regard but, unfortunately, had not really reached the stage of general applicability. While Bush was not able to furnish any further advice, he pointed to a general conceptual difference between Zworykin's methods of prediction and his own imagination of how to realize a prediction apparatus.

> However, the handling of the equations of dynamic meteorology, while undoubtly fruitful, is only one approach to the subject. As I understand it your approach is quite different. It consists in essence of the construction of a model, using primarily electric analogs, which is so flexible that it can be caused empirically to behave in the same way that measurements indicate the variables of weather performance to behave over a particular region.
>
> (Bush 1945)

This short correspondence already shows the general ambiguity at the end of World War II on how mechanized and electrified weather forecasting would have to be designed. Vladimir Zworykin's approach, although also factoring in the solution of "dynamical equations" through extrapolation techniques, corresponds more to a simulation in the traditional sense of "imitating" or "reproducing" a certain behavior by means of a functionally analog model: an automatic plotting board, by which the electrical components "behave in the same way" as the measurement variables of the actual weather "perform." Or as Jule Charney described this "analog simulation" method much later: "At the time I didn't take the proposal seriously, because he [Zworykin] actually expected to reproduce the weather in a cathode ray tube" (Platzman 1987: 54). Zworykin actually envisioned an electrical representation of the weather, extrapolating a future state from the current state by using statistical reference values from the past. These past states, in turn, would be provided by automatic selection procedures "of the IBM type," that is, punch card sorting machines. Bush, in turn, had an alternative understanding of the meteorological problem. He saw the crux and thus also the whole difficulty of accurate weather forecasts in the handling of the primitive equations of fluid motion, for which, as yet, no satisfactory automated solution had been found. While the servomechanical principle of the differential analyzer failed to meet the needs of modeling the hydro- and thermodynamic evolution of weather systems, the more promising attempt of photoelectrically scanning an integral with the optical or cinema integraph also seemed to go nowhere (Bush 1936: 659–660).

Not least because of Bush's clairvoyant remarks, Zworykin prepared a revised version of his pamphlet, adding an introduction on the new availability of massive, worldwide weather information as well as a brief bibliography of meteorological literature he had consulted and distributed the paper to further "foresighted men." Apparently, the forecasters of the Weather Bureau did not yet fall into this category. But a certain John von Neumann at the nearby Institute for Advanced Study did.

Von Neumann had been on the core staff of the IAS since its founding in 1933. During the war, he had already dealt with several hydrodynamic problems, which essentially revolved around the problem of shock wave propagation, an essential complication in the design of the nuclear bomb. Since early 1944 most of his work at the IAS entailed directing a project for the "Applied Mathematics Panel" of Bush's OSRD. "The object of this project was to carry out calculations on various aerodynamical questions of military importance, and also to develop new computing methods which are likely to be useful in this field" (Neumann 1945a).[8] He developed a marked interest in the numerical methods of fluid mechanics and their approximate integration by means of computing machines and, consequently, the logical design of these computing machines themselves. Since spring 1945, von Neumann had been drumming up support for the construction of an "experimental computer" at his home institution, pointing to the fact that the US government might be willing to reallocate funds for physics and engineering from war research to academic institutions.[9]

"Experimental" is a key word here, and is to be understood in a twofold sense. On the one hand, a machine was to be designed that could overcome the stagnation in the theoretical treatment of certain problems in mathematical physics that escaped any analytical solution, and thus help to replace the hitherto empirical approaches in attacking them. This effected problems of aero- and hydrodynamics, as well as issues of celestial mechanics, quantum theory, optics, electrodynamics, and also certain fields within economic or statistical theory. "The problems are so varied, and in many cases their details as yet unpredictable, it would be unwise to build now a 'one-purpose' machine for any one problem or closely circumscribed group of problems" wrote von Neumann. "An 'all-purpose' device should be the aim" (Neumann 1945b). However, the operating principles and machinery of electronic computing itself should be investigated experimentally, leading to a more conceptual design approach for a general-purpose computer.

> I wish to emphasize that our idea is to build this device as a research tool in order to render an effective study of the problems outlined in (1) to (3) [new computing methods, its arithmetical, control, memory, input and output 'organs' as well as the general theory and philosophy of the use of such devices] possible ... [W]e wish to discover the best methods of computing, of complementing experimentation by computation, and of controlling computing and other processes.
>
> (Neumann 1945d: 2)

Until now, it was clear to Neumann only that the special class of nonlinear problems called for a computer design that was both digital and electronic. Analog representations of decimal places by relays would be just as inadequate for the precision requirements of numerical methods as were the switching times required for this purpose.

> [T]hose problems which offer the most interesting and important uses for future computing machines, and in particular the device which we plan, cannot be done at all on any differential analyzer with practicable characteristics.
>
> (Neumann 1945c)

Thus, the task was compounded. Immediately following the authorization of funds for von Neumann's plans for an "Electronic Computer Project" on 19 October 1945, Zworykin's proposal landed on his desk (IAS 1945: 9).[10] Von Neumann, who himself had, at an earlier occasion, brought up the example of dynamic meteorology as a possible field of application for his computer (Neumann 1945b: 5), was infected by Zworykin's enthusiasm. In his reply of October 24, he went straight into the discussion:

> I think that the mathematical problem of predicting weather is one which can be tackled, and should be tackled. It will require very extensive computing, but the equipment to do this is now becoming available or can

be developed. Clearly the problem can be attacked on various levels of abstraction: anywhere between the purely sorting approach which compares present weather maps with past ones and attempts to establish the closest analogous past situation in order to extrapolate by past experience; and the entirely aerodynamical one which would aim to compute the movements of air masses starting from the present distribution of pressures, temperatures, wind velocities, humidity, and the states of radiation, reflection, and absorption.

(Neumann 1945e: 1)

Once again, another methodological difference was drawn here in all clarity. Von Neumann distinguished between the comparison with "analogous situations of the past," that is, the generic practice of veteran meteorologists and climatologists at the Weather Bureau and elsewhere that maps the data, and the "aerodynamic" method that tackles the problem of prediction by directly calculating a future state on the basis of given diagnostic variables. There is no need to say which methodological approach Neumann preferred. As a mathematician and theoretical physicist, and especially given his experience with the computerized calculation of shock wave propagation as part of his participation in the Manhattan Project, he regarded the fluid mechanical approach as the more interesting one. Still, he acknowledged serious difficulties:

If it were not for the considerable uncertainties of turbulent heat transfer, even the latter purely-theoretical approach would be feasible with electronic computing that might be available within the next few years. At any rate this aspect of the problem, together with the important questions of turbulence with which it is connected, should certainly be investigated.

(Neumann 1945e: 1)

As an intermediate step, he continued, one would probably have to apply statistical and empirical description methods to master turbulent energy transfer, yet another instance to use the full capabilities of (digital and analog) computer equipment. Once the technical predictability is finally given, the next target then would be weather modification:

I agree with you completely that once the methods of prediction are sufficiently advanced the immediately following step should be prediction from hypothetical situations. In other words: exploring the consequences of various controllable changes ... of suitable atmospheric phenomena which can be brought about artificially.

(Neumann 1945e: 2)

Apparently, von Neumann shares the assumption underlying Zworykin's proposal, namely the possibility to intervene by applying amounts of energy only a fraction the size of the actual phenomenon.

I agree with you that our present inability to influence the weather is not due to the fact that the energies involved are too great, since the most conspicuous meteorological phenomena originate in unstable or metastable situations which can be controlled, or at least directed.

(Neumann 1945e: 2)

One could debate whether such lines present von Neumann as a true "climate engineer," and how such thoughts might be connected to his later fatalism on the power of humankind over weather and climate processes (Fleming 2010: 190–191; Kwa 2001: 141–142).[11] However, for the purpose of this paper it is important to note the manifestation of a third perspective on the problem of automatic weather forecasting, namely the one that would prevail. Von Neumann wanted to achieve progress on an entire class of universal physical-mathematical problems with the aid of new numerical methods, and concurrently develop the basic design of their automated solution in a conceptual as well as experimental fashion. Here, dynamic meteorology comes in quite handy, as it appears to be a welcome scientific field for testing both the relevance and the power of this approach.

The eventual result is well known: the digital stored-program computer, generally known as the "Princeton" or "von Neumann architecture," in which both instructions and data are stored in the same memory. Initially built on the grounds of the IAS as the material realization of the Meteorological Computing Project, derivatives of this machine were widely distributed to other institutions such as the RAND Corporation (JOHNNIAC), the Los Alamos National Laboratories (MANIAC 1), and the Swedish Board for Computing Machinery (BESK), on which the first operational numerical weather forecast was achieved in 1954.

A third reaction: Meteorologists

Let's summarize the intermediate results for the moment. We have an electronics engineer who sees the vaguely recognizable potential of computational electronics to speed up statistical and diagrammatical analysis and hereby more or less equates signaling with real geoengineering. We have an administrator of the technical sciences, who by way of his own historical involvement and his experience with the design of analog computing machines, highlights their specific inadequacy for the purpose of fluid dynamical problems. And we have a mathematician who seeks to pave the way out of this dilemma in a grand universalist program for which he welcomes weather forecasting as an ideal application. Put emphatically, and in a slightly exaggerated way: Zworykin wants to switch the weather, Bush is looking for a special apparatus approach to the forecasting problem, and von Neumann wants to analyze the operating principles of a universal scientific computing machine by way of tackling the fluid dynamics of atmospheric motion.

This reflects not only three different epistemological approaches to the creation of future prediction techniques and apparatus, it also makes clear that all three actors are merely outsiders to the meteorological profession. The question arises as to who actually contacted and consulted the real meteorologists from the Weather Bureau? And in which way did all these considerations by outsiders have an impact on forecasting operations?

To begin with, the Weather Bureau was very confused. Under "mechanical handling" the Weather Bureau personnel understood technical assistance to accomplish their everyday data processing routines. That is why the Weather Bureau is called the Weather "Bureau". It had to manage and process the observational variables on a daily basis: for government, the business community, and the farmers. The swelling of the meteorological and especially aerological data deluge over the previous decades had already brought some familiarity with telex machines, desk calculators, and, much later than in the European meteorological services (Nebeker 1995; Kistermann 1999), with punch card equipment of any kind, that is, with machines to duplicate, check, and translate punched cards. In fact, the increased use of mechanical computing and recording devices in the interwar period—regular, commercially distributed instruments like Felt's Comptometer, the Brunsviga and the Mercedes-Euclid calculating machine—had led to an increased focus on the statistical, that is, climatological aspects of meteorology (Nebeker 1995: 93–94). The massive expansion and increased density of the observation network over the course of World War II, accomplished by ground troops stationed on all continents and by warships sailing the high seas, completely overwhelmed the data and analysis system of the Weather Bureau. Proposals such as Zworykin's automated handling of punch-coded data and its subsequent statistical analysis seemed quite promising to the Bureau. "This is a startling but noteworthy proposal," director Reichelderfer stated in a circular to his close associates when he finally, after several weeks, received a copy of Zworykin's paper (Reichelderfer 1945).[12]

Another indication of the welcoming reaction is the response to Zworykin's proposal by oceanographer and meteorologist Athelstan Spilhaus. At the time, Spilhaus was researcher at the Camp Evans Weather Laboratories of the US Army Air Force R&D branch in Belmar, New Jersey, and hence much acquainted with the daily grind of weather forecasting. He wrote:

> The application of electrical and electronic computers to meteorological prediction may be regarded as merely another step (but potentially a great one) that has already started. Prior to the war, the factor of personal experience and local knowledge was a great one in the make-up of a successful forecaster. His experience often resolved into a remembrance of a situation similar to the one whose development he was currently called upon to predict. It was a short step to the process of classifying map types and subsequently to arranging these so that matching could be accomplished automatically by punched-card index machine rapid reference methods. The use of the electrical computers is as logical a step and coupled with

type matching equipment will be a tremendously powerful tool as the methods may be continually checked by experiments on past sequences.
(Spilhaus 1945)

Filing, sorting, and comparing the stocks of meteorological and climatological memory: such was the future task of electronic labor slaves as envisioned by meteorological practitioners. "Experimental" is to be seen here as exercises in diagrammatical pattern recognition, matching current maps to their historical archive. Integrating the equations necessary to model the process evolution of a fluid-dynamical space-time system was exceptionally alien to them. Indeed, as far as the theoretical reformulation of meteorology itself was concerned, the Bureau seemed to have been largely unprepared. Not a single project employed at the end of World War II dealt with general physical questions, let alone with the numerical solution of the nonlinear equations of atmospheric dynamics (Harper 2008: 95–96). As for von Neumann's ambitious high-speed calculation, it remained empty-handed. There was practically nothing to calculate!

The starting point of numerical meteorology at the end of 1945 could hardly have been more disparate: Zworykin wanted to bring the weather, von Neumann electronic computing, and the Weather Bureau its free-floating data under control. As all three of them gathered at a small ad hoc conference at the headquarters of the Weather Bureau on 9 January 1946 (Reichelderfer 1945),[13] the situation gradually began to unfold. John von Neumann broadened the already approved "Electronic Computer Project" to include a "Meteorology Project," effectively turning the entire endeavor into a combined "Meteorological Computing Project" to "study the application of electronic computation to dynamic meteorology" in order to make feasible "short and long range weather forecasting, or even control in a sense" (Dyreson 1946: 1). Vladimir Zworykin's R&D laboratories at RCA started to develop the storage tubes required to build the computer. And the Weather Bureau regularly sent its chief researcher, Harry Wexler, to the Princeton area to report on progress. Only when the vodka flowed a little more turbulently were speculations about man-made oil spills and deflected hurricanes ventured again.

Conclusion

From this time, the "epistemo-technical" (Hörl 2008) dance between mathematical theory, numerical experimentation, and computer design has taken shape continuously, substantially transforming meteorological practice and the scientific culture around it. This process, while repeatedly interspersed with all kinds of serious setbacks of a technical, institutional, or social nature (as expounded by Harper 2008), gradually established a new stability on which the numerical treatment of atmospheric dynamics rests, basically until today. Eric Winsberg, in keeping with Ian Hacking and Peter Galison, stated that "various experimental techniques and instruments develop a tradition that gives them their own internal stability, or, put most provocatively, that 'experiments have a life of their own'" (Winsberg 2003: 121).

However, this historical miniature refers to the brief moment in which this tradition was just about to emerge, hence a moment of utter epistemic instability. Analyzing the correspondence among these three proponents of distinctive cultures of prediction, each with their own visions of the relationships between the material cultures of computation, its institutional formations and wider political purpose, it seems apparent that the future was wide open and negotiable and that some kind of year-zero mindset reigned in 1945, in which everything seemed possible and nothing too far-fetched to attract the attention of the scientific and engineering communities as well as research grants. It was certainly a historical instance of open trajectories, albeit not beyond theoretical possibilities and pragmatic necessities. It was a moment highly vulnerable to the contingencies of scientific progress. It is hardly idle to speculate about what would have happened to weather prediction, climate research, or the whole of today's computational sciences if, say, an influential science policy actor like John von Neumann had not picked meteorology as a key discipline to apply for government funds to probe the potential of binary electronic computing, an "enterprise clearly [...] a gamble," albeit "a reasonable gamble" (Neumann 1945f).

On 30 January 1947, the Princeton duo Zworykin and von Neumann appeared consecutively before a joint session of the American Meteorological Society and the Institute for Aeronautical Sciences in Hotel Astor, New York City. Zworykin, drawing all the attention of attending journalists, once again repeated his vision of drastic weather interventions aided by exact and high-speed computing methods (Anonymous 1947a, 1947b). John von Neumann had another interesting outlook. After explaining the meteorological prediction problem and the need for an "all-purpose" digital machine to tackle it, he went on to say:

> Special, specifically meteorological, electronic machines should be developed thereafter. These machines should be conceived as combining in an appropriate manner digital (that is, arithmetical) and analog (that is, small scale model) components [...] alternating for various phases of the total problem as they are best suited.
>
> (Neumann 1947)

It appears that, in the long run, von Neumann envisioned quite another type of machine to aid both forecasting and "mathematical experimentation" for "hypothetic situations" (Neumann 1947), that is, simulation: a hybrid apparatus combining digital and analog methods. No doubt, even John von Neumann would be bewildered to see today's fully digitized infrastructure of meteorology and the climate sciences.

Notes

1   Most notably, Kristine Harper (2008) has described, in full chronological detail, the individual and institutional context within which the rise of NWP is to be situated. See also Lynch (2006). Earlier, but still valuable accounts include Nebeker (1995) and William Aspray (1990), who devoted a full chapter to "The Origins of Numerical Meteorology." Fully aware of their own place in history, the peculiarity of the transformation was

celebrated by their proponents early on, e.g. by Thompson (1983, originally published in 1976). On the general transformation of the sciences toward simulation see Gramelsberger (2011).

2 In the context of this volume it is interesting to note that even the study of history is evaluated to become a pattern-analytical and predictive science (Turchin 2008).

3 Well known is Norbert Wiener's self-proclaimed history of cybernetics, in which his design of an automated anti-aircraft predictor plays the central role (Wiener 1985a, originally published in 1958). However, he also makes clear in this essay how much both his studies of mechanization for solving partial differential equations (PDE) and his long-standing interest in (statistical methods for) predicting the weather influenced his thinking on cybernetics and the theory of communication. Indeed, there is a fourth version of weather prediction to be portrayed here, namely Wiener's harmonic analysis, ergodic theory and his own design of an apparatus for the solution of PDEs (Wiener 1985b, originally written in 1940). But as Wiener's work would require a longer introduction and this paper wants to avoid repeating the classical comparison of the different (but sometimes also very similar) visions of von Neumann and Wiener, this eminent figure is left out of the discussion here.

4 Demonstrating the "longue durée" inherent in the conceptual appropriation of atmospheric motion is one of the central aims of my current book project. In this, I show how the encoded heuristics are founded on a long descent of technical apprehensions of flow, and that perceived "scientific revolutions" are mere technological shifts in a general epistemic frameworks based on a long tradition of comprehending Earth as a fluid-dynamical system.

5 With this letter Condon, director of the National Bureau of Standards, had sent two copies of Zworykin's proposal to Reichelderfer. A side note remarks that one of them was filed by Wexler on 18 January (Condon, 1946).

6 Zworykin summarized the idea as follows: "Thus, while the energy finally released may be enormous, that required to trigger the release may be quite modest. Furthermore, the magnitude of the triggering energy required will greatly depend on the time and place at which it is applied" (Zworykin, 1947: 28); Elsewhere, Zworykin calculated the amount of energy in a historic hurricane over Puerto Rico in 1899 as 20,000 times that of the atomic bomb detonated over Hiroshima only a month earlier (Zworykin, 1945a: 8).

7 The one contact with weather forecasting that Bush had to deal with primarily during his time as president of the Carnegie Institution for Science (1938–55) was to spur the much delayed write-up (and chart drawing) of the third and final volume of Vilhelm Bjerknes's magnum opus *Dynamical Meteorology and Hydrography*, which had been funded by Carnegie since 1906 (Bjerknes 1910, 1911, the third volume never got published). See his great volume of correspondence on this matter in Carnegie Institution of Washington, Administration Files, Bjerknes, V., 3, 32. Being the most proficient and reliable source for judging any development in dynamic meteorology during that time, Carl-Gustav Rossby advised him to further delay publication until he had the opportunity to include the wealth of "modern ideas" currently under development (Rossby 1940). In fact, Kristine Harper regards Rossby—and not the ubiquitous von Neumann—as the true hero of the entire story (Harper 2004, 2008).

8 The contract with the Applied Mathematics Panel, headed by Warren Weaver, states that the emphasis of the project was on "a mathematical theory of gas dynamics" (IAS Archives, Aydelotte Papers, Government Contracts). Despite the divergence in scientific nomenclature, it is clear that von Neumann mainly had to deal with those fundamental hydrodynamical problems affecting the Manhattan Project's research on high explosives (Neumann 1995a, originally written in 1943; Winsberg 2003: 113–114.).

9 Quite instructive in this regard are the minutes of one of the meetings held by all professors of the IAS School of Mathematics. After a longer discussion on how to attract leading figures of the just winding-down war research effort to the IAS—a discussion that

included Albert Einstein's repeated warnings against any involvement in "preventive" war research—von Neumann raised the opportunity for the IAS to play a "directing role" in exploring the potential of "automatic computing" (IAS Archives, General, Hel-Hiz).

10 Coincidence has it, that in the same meeting it was also approved that a member of IAS's School of Humanistic Studies, the palaeographer Elias Avery Lowe, was allowed to sell his house to Vladimir Zworykin for $30,000, effectively turning Zworykin into a resident on IAS property (Stern 1964: 608).

11 John von Neumann's famous *Can we survive technology?* of 1955 is still a fascinating piece, especially when looking back at this time as the start of the "Great Acceleration" (e.g. Steffen et al. 2011: 849–852). Von Neumann writes: "the crisis is due to the rapidity of progress, to the probable further acceleration thereof, and to the reaching of certain critical relationships. Specifically, the effects that we are now beginning to produce are of the same order of magnitude as that of 'the great globe itself.' Indeed, they affect the Earth as an entity. Hence further acceleration can no longer be absorbed as in the past by an extension of the area of operations" (Neumann 1995b: 672).

12 Having received two copies via Edward Condon (s. fn. 4) on 27 Nov, he immediately invited Zworykin to the Weather Bureau to acquaint himself with the daily operations. Zworykin happily accepted, further requesting to bring two engineers with him (Zworykin 1945c).

13 In fact, the significance of this meeting for the entire history of numerical weather prediction is mirrored by Kristine Harper's choice to start her book with this gathering of von Neumann, Zworykin and the Weather Bureau (Harper 2008: 1).

References

Anonymous (1947a) "Can storms be controlled? Dr. V. K. Zworykin discusses possibilities of weather control and significance of the electronic computer," *Radio Age* 6(3), pp. 9–10.

Anonymous (1947b) "Storm Prevention Seen by Scientist," *New York Times*, 31 January 1947.

Aspray, W. (1990) *John von Neumann and the Origins of Modern Computing*, Cambridge, MA: MIT Press.

Bjerknes, V. (1910) *Dynamical Meteorology and Hydrography*, (vol. 1: Statics), Washington, DC: Carnegie Institution of Washington.

Bjerknes, V. (1911) *Dynamical Meteorology and Hydrography*, (vol. 2: Kinematics), Washington, DC: Carnegie Institution of Washington.

Bush, V. (1936) "Instrumental analysis," *Bulletin of the American Mathematical Society* 42(10), pp. 649–669.

Bush, V. (1945) *Letter to Vladimir Zworykin, 18 Sep 1945*, Library of Congress (LoC), Vannevar Bush Papers, 122.

Condon, E. U. (1945) *Letter to Francis W. Reichelderfer, 26 Nov 1945*, LoC, Harry Wexler Papers, 2, 1946, 7.

Dyreson, A. (1946) *Justification Memorandum for Navy Contract, PD#EN1-22/00028, 6 June 1946*, LoC, John von Neumann Papers, 15, 6.

Edwards, P. (2010) *A Vast Machine: Computer Models, Climate Data, and the Politics of Global Warming*, Cambridge, MA: MIT Press.

Fleming, J. R. (2010) *Fixing the Sky: The Checkered History of Weather and Climate Control*, New York: Columbia University Press.

Gramelsberger, G. (ed., 2011) *From Science to Computational Sciences. Studies in the History of Computing and its Influence on Today's Sciences*, Zurich and Berlin: Diaphanes.

Guillemot, H. (2010) "Connections between simulations and observation in climate computer modeling. Scientist's practices and 'bottom-up epistemology' lessons," *Studies in History and Philosophy of Modern Physics* 41(3), pp. 242–252.

Harper, K. C. (2004) "The Scandinavian Tag-Team: Providers of atmospheric reality to numerical weather prediction efforts in the United States (1948–1955)," *Proceedings of the International Commission on History of Meteorology* 1(1), pp. 84–91.

Harper, K. C. (2008) *Weather by the Numbers: The Genesis of Modern Meteorology*, Cambridge, MA: MIT Press.

Heymann, M. (2010) "Understanding and misunderstanding computer simulation: The case of atmospheric and climate science—An introduction," *Studies in History and Philosophy of Modern Physics* 41(3), pp. 193–200.

Hörl. E. (2008) "Die epistemotechnische Transformation. Metatechnische Reflexionen über das Problem des Wissens im Zeitalter der Simulation," A. Gleiniger and G. Vrachliotis (eds.) *Simulation. Präsentationstechnik und Erkenntnisinstrument*, Basel, Switzerland: de Gruyter, pp. 93–106.

IAS Institute for Advanced Study (1945) *Minutes of Regular Meeting of the Board of Trustees, 19 Oct 1945*, IAS Archives.

Kistermann, F. W. (1999) "Leo Wenzel Pollak (1888–1964): Czechoslovakian Pioneer in Scientific Data Processing," *IEEE Annals of the History of Computing* 21, pp. 62–68.

Kwa, C. (2001) "The Rise and Fall of Weather Modification," C. A. Miller and P. N. Edwards (eds.) *Changing the Atmosphere: Expert Knowledge and Global Environmental Governance*, Cambridge, MA: MIT Press, pp. 135–165.

Lewis J. M. (2000) "Clarifying the dynamics of the general circulation: Phillips's 1956 experiment," D.A. Randall (ed.) *General Circulation Model Development: Past, Present, and Future*, San Diego, CA: Academic Press, pp. 91–125.

Lynch, P. (2006) *The Emergence of Numerical Weather Prediction: Richardson's Dream*, Cambridge, MA: Cambridge University Press.

Lynch, P. (2008) "The ENIAC forecasts. A recreation," *Bulletin of the American Meteorological Society* 89, pp. 45–55.

Nebeker, F. (1995) *Calculating the Weather: Meteorology in the 20th Century*, San Diego, CA: Academic Press.

Neumann, J. von (1945a) *Letter Summarizing his War Activities to Frank Aydelotte (Director IAS), 11 Oct 1945*, IAS Archives, Faculty, von Neumann, 1941–1951.

Neumann, J. von (1945b) *Memo to Frank Aydelotte, 5 Sep 1945*, LoC, John von Neumann Papers, 12, 1, 1.

Neumann, J. von (1945c) *Letter to Frank Aydelotte, 1 Nov 1945*, LoC, John von Neumann Papers, 12, 1.

Neumann, J. von (1945d) *Letter to R. W. Brown (Office or Research and Invention, Navy Department), 28 Nov 1945*, LoC, John von Neumann Papers, 15, 2.

Neumann, J. von (1945e) *Letter to Vladimir Zworykin, 24 Oct 1945*, LoC, Harry Wexler Papers, 18, 1.

Neumann, J. von (1945f) *Letter to Harry Wexler, 28 Jun 1946*, LoC, Harry Wexler Papers, 2, 1946, 5

Neumann. J. von (1947) "The future role of rapid computing in meteorology," *Aeronautical Engineering Review* 6, p. 30.

Neumann, J. von (1960) "Some remarks on the problem of forecasting climatic fluctuations," R. L. Pfeffer (ed.) *Dynamics of Climate: Proceedings of a Conference on the Application of Numerical Integration Techniques to the Problem of the General Circulation, held Oct. 26–28, 1955*, Oxford: Pergamon Press, pp. 9–11.

Neumann, J. von (1995a) "Theory of shock waves," F. Brody and T. Vamos (eds.) *The Neumann Compendium*, Singapore: World Scientific, pp. 378–402.

Neumann, J. von (1995b): "Can we survive technology?" F. Brody and T. Vamos (eds.) *The Neumann Compendium*, Singapore: World Scientific, pp. 658–673.

Platzman, G. W. (1979) "The ENIAC computations of 1950—gateway to numerical weather prediction," *Bulletin of the American Meteorological Society* 60, pp. 302–312.

Platzman, G. W. (1987) *Conversations with Jule Charney, 1987*, National Centre for Atmospheric Research (NCAR) Archives.

Reichelderfer, F. W. (1945) *Memo 29. Dec 1945*, LoC, Harry Wexler Papers, 2, 1945, 2.

Rosol, C. (2007) *RFID. Vom Ursprung einer (all)gegenwärtigen Kulturtechnologie*, Berlin, Germany: Kadmos.

Rossby C.-G. (1940) *Letter to Vannevar Bush, May 15 1940*, Carnegie Institution of Washington, Administration Files, Bjerknes, V, 3, 32.

Serres, M. (2003) "La simulation, technique nouvelle, ancienne tradition," *Clefs CEA* (Commissariat à l'Energie Atomique) 47, pp. 2–5.

Spilhaus, A. (1945) *Comments on Weather Proposal, 6 Nov 1945*, Harry Wexler Papers, 18.

Steffen, W., J. Grinevald, P. Crutzen and J. McNeil (2011) "The Anthropocene: conceptual and historical perspectives," *Philosophical Transactions of the Royal Society A*, 369, pp. 842–867.

Stern, B. M. (1964) *History of the Institute for Advanced Study 1930–1950*, typoscript, IAS Archives, 608.

Thompson, P. D. (1983) "A history of numerical weather prediction in the United States," *Bulletin of the American Meteorological Society* 64(7), pp. 755–769.

Turchin, P. (2008) "Arise 'cliodynamics'," *Nature* 454, pp. 34–35.

Wexler, H. (1946) *Letter to Francis W. Reichelderfer, 18 Oct. 1946*, LoC, Harry Wexler Papers, 2, Wexler, 1946, 7.

Wiener, N. (1956) "Nonlinear Prediction and Dynamics," J. Neyman (ed.) *Proceedings of the Third Berkeley Symposium on Mathematical Statistics and Probability*, (vol. 3), Berkeley, CA: University of California Press, pp. 247–252.

Wiener, N. (1985a) "My connection with cybernetics. Its origins and its future," P. Masani (ed.) *Collected Works with Commentaries* 4, pp. 107–120.

Wiener, N. (1985b) "Memorandum on the mechanical solution of partial differential equations," P. Masani (ed.) *Collected Works with Commentaries* 4, pp. 125–134.

Winsberg, E. (2003) "Simulated experiments: methodology for a virtual world," *Philosophy of Science* 70, pp. 105–125.

Zworykin, V. K. (1945a) *Outline of Weather Proposal, Sep 1945*, (attachment to a letter to Vannevar Bush of 14 Sep. 1945), LoC, Vannevar Bush Papers, 122.

Zworykin, V. K. (1945b) *Outline of Weather Proposal, Oct. 1945*, LoC, Harry Wexler Papers, 18, 1.

Zworykin, V. K. (1945c) *Letter to Francis W. Reichelderfer, 6 Dec 1945*, LoC, Harry Wexler Papers, 2, 1946, 7.

Zworykin, V. K. (1947) "Electronic Weather," *Yankee Magazine*, pp. 27–28.

5   A new climate

Hubert H. Lamb and boundary work at the UK Meteorological Office

Janet Martin-Nielsen

We are living in a time when the glamour of the much more expensive work of the mathematical modeling laboratories, and the tempting prospect of theoretical predictions, are stealing the limelight.

(Lamb 1997: 203)

No scientific understanding can be built up without first observing and recording the phenomena and processes that have to be explained. Reconstruction of the past record of weather and climate is necessary in order to provide climatology with its observational base.

(Lamb, n.d.1, Mapping Historical Weather and Past Climates: 1–2)

Introduction

It is clear that computer modeling represents the leading method of climate research and prediction today, and has enjoyed this status since the 1970s (Weart 2010; Edwards 2011; Guillemot 2007). Indeed, as Environment Canada geographer, Stewart Cohen, and his colleagues remarked more than a decade ago, the success of modeling "has served to marginalize other, less reductionist ways of understanding global climate change, most notably paleo-climatic and other analog methods" (Cohen et al. 1998: 345).[1] The emergence of modeling as the predominant research strategy for climate, however, was neither straightforward nor predetermined: rather, models arose from "a competition between different knowledge claims and epistemic standards and attained hegemonic status within a diversity of knowledge cultures" (Heymann and Martin-Nielsen 2013: 1). The postwar decades saw an abundance of boundary work as the discourses, epistemic standards, research cultures, and scientific-political-social legitimacy of new ways of thinking about climate were worked out (Gieryn 1983; Gieryn 1999). This chapter looks at these broad ideas through the lens of English meteorologist and climatologist, Hubert Horace Lamb (1913–1997), and the UK's national weather service, the Meteorological Office.

A changing climate

In July 1956, the Royal Meteorological Society announced the second Napier Shaw Memorial Prize essay competition, "open to anyone without restriction of nationality," with a prize of £100 (in 2015 figures, approximately £1,700) (The Napier Shaw Memorial Prize,1956: 383). "The essay may deal with climatic variation on any time scale or scales, [the prize announcement read:] A critical analysis and appraisal of the evidence is sought concerning a large part of the earth's surface and preferably on a global basis" (ibid.). The due date was 1 January 1959, two and a half years away. Named for Sir Napier Shaw (1854–1945), the first director of the Meteorological Office and the first professor of meteorology at Imperial College London, the prize commemorated the birth centenary of the dean of turn-of-the-century English meteorology. Moreover, the topic chosen by the prize committee—climatic variation—was representative of the changing conception of climate at the time, and spoke directly to the heart of a changing discipline.

In the middle of the twentieth century, the concept of climate shifted from a stable, geographical concept to a dynamic concept linked not to region but to weather patterns (Heymann 2009; Heymann 2010). This marked a significant change from the late nineteenth century, when it was widely agreed that climate was a stable phenomenon that had experienced no significant changes in thousands of years.[2] Within this "classical climatology," climate was typically seen as an ensemble of data about "normal" weather phenomena (e.g. temperature, wind and precipitation) for a given region, liable only to limited variability: climate was a static regional and spatial concept, and climatology was a geographical science. According to English meteorologist and climatologist, Hubert Horace Lamb, in 1964:

> This view of climate as effectively constant, undergoing at most cyclical changes of no great importance, was taught to most of my generation in school ... It was believed that you only had to average data over a long enough period of time to arrive at a figure to which the climate would always return.
>
> (Lamb 1964: 171)

Growing recognition of warming trends in the early decades of the twentieth century, however, contradicted the climatic stability at the heart of classical climatology. "Interest in the subject of climatic change was aroused once the considerable warming of our climate in most seasons of the year from the 1890s to the 1930s and '40s became obvious to all," Lamb reported to the British Association for the Advancement of Science in 1964 (Lamb 1964: 171–172).[3] Swedish glaciologist Hans W. Ahlmann, Swedish meteorologist Carl-Gustaf Rossby, and German climatologist Helmut E. Landsberg, among others, also spoke out about this warming, with Rossby even taking his concerns to the Pentagon during a visit to the United States in the summer of 1947

(Doel 2009: 143). And this new view of climate was far from constrained to the scientific community: popular publications from *Time* magazine to *The New York Times* reported on the scientific discourse and offered graphic and sometimes alarming descriptions of what a warming climate might mean for the planet.[4] "Subzero temperatures occur only half as frequently in northern cities as they did 75 years ago," asserted American science journalist, Waldemar Kaempffert, in *The New York Times* in 1952:

> Eskimos are catching and eating cod, a fish that they never saw before 1900. Greenland's ice is melting, and in the process ruins of ancient farmhouses have been uncovered. Ships ply the White Sea and the Gulf of Bothnia three or four weeks longer in winter than they used to. [And in] Iceland and the higher latitudes of Norway farmers are growing barley in soil that used to be frozen seven months in the year.
>
> (Kaempffert 1952)

The idea that "if only the observed values of the meteorological element were averaged over a sufficiently long period of years the result could be defined as a 'normal' value, to which the element would always tend to return," as Lamb put it, soon came to be seen as "mistaken"—and by the mid-to-late 1950s, the idea of climatic consistency had been dealt a blow, to the extent that there was even speculation about an ice-free Arctic Ocean within decades (Lamb and Johnson 1959: 102–103).

As climatic consistency began to be doubted and then rejected, decades of working assumptions and methodologies were questioned and the concept of climatic prediction was newly open. Up to the 1950s, historian Spencer R. Weart writes,

> the usual assumption was that climate conditions of recent decades could be used to predict what to expect. If the region was subject to droughts and other transient changes, it was hoped that they could someday be predicted through analysis of regular cycles, linked for example to sunspots.
>
> (1998: 99)

But as the view that climate could change on human timescales prevailed, this methodology was discredited. "Tables of climatic statistics could no longer be used with confidence as a guide to the future—or at least, not without asking which past years' data they comprised and what reason there was for supposing that these were the best guide to the planning period," Lamb explained in *Nature* (Lamb 1969a: 1209). New frameworks, new methodologies, and new ways of thinking were needed.

At the Meteorological Office, the emergence of a dynamic conception of climate forced both leaders and scientists to reconsider working assumptions, research programs, and future plans.[5] Epistemic authority, knowledge production, professional legitimacy, and research cultures all hung in the balance.

This chapter deals with one manifestation of this situation: the conflicting philosophical approaches to climate research, which arose at the Meteorological Office in the mid-to-late 1960s—a conflict that ultimately resulted in Lamb's departure from the Office and his establishment of the Climatic Research Unit at the University of East Anglia.

Hubert H. Lamb and the Royal Meteorological Society's essay prize

At the time of the Royal Meteorological Society's essay competition announcement, Lamb had been working at the Meteorological Office for nearly twenty years. He had been stationed in Scotland and on a secondment in Ireland, as well as abroad in the Antarctic waters, in Malta and in Libya—a variety of postings, which had built in Lamb an interest in large-scale mappings of the atmospheric circulation.[6]

In mid-1956, he was sent back to the UK, to the Office's climatology section at Harrow, in northwest London. His primary task was to respond to climatic inquiries from overseas, but he was drawn to the archives stored two floors below ground at Harrow—archives which were arguably "the richest resource anywhere in the world of past meteorological observations" (Lamb 1997: 179). With weather diaries, maps, almanacs, and quantitative observations spanning centuries, these archives—dusty, disorganized and of little interest to anyone else at the Meteorological Office—kindled Lamb's long-standing interest in history, gave him material for his essay competition entry, and would come to shape his views on climate and climate research for decades to come.

When he heard about the essay prize, Lamb immediately decided that he "had to have a try" (Lamb 1997: 181). After receiving permission from Oliver Graham Sutton, director of the Meteorological Office from 1954 until 1965, to focus his research on climatic variation, Lamb spent, in his words,

> many happy hours of intense work in the archives, wonderfully free from interruptions there two floors below the ground, discovering the great wealth of weather observation records that had been collected in the nineteenth and early twentieth centuries, and bound together in handsome volumes here, in the headquarters of the British Meteorological Service.
>
> (Lamb 1997: 181–182)

Lamb began digging through the Harrow archives with the intention of constructing global pressure maps stretching as far back in time as possible—a task he hoped would catch the attention of the prize committee. Lamb spent two and a half years immersed in the untapped archives of climate data, emerging in 1959 with an essay entitled "Climatic Variation and Observed Changes in the General Circulation" (Lamb and Johnson 1959).

The essay provides an early enunciation of Lamb's philosophical approach to climatology. The impact of this philosophy would reach its apex in 1971, when Lamb quit the Meteorological Office after thirty-six years of service—a move prompted by his frustration with the Office's focus on numerical models and its exclusion of his preferred historically-based research methodology. In this light, the pressure map essay enables us to better understand both Lamb's early thinking about climate and the boundary work which took place in the Meteorological Office as numerical models gained increasing hegemony in the mid-to-late 1960s.

The records in the Harrow archives contained data from sources as diverse and far-flung as the Societas Meteorologica Palatina's late eighteenth-century observation stations stretching from the Urals to North America (see also Gramelsberger in this volume), British naval ships on mid- nineteenth-century over-winter expeditions in the high Arctic, and E. A. Holyoke, an octogenarian medical doctor and amateur weather observer in Salem, Massachusetts, whose daily weather data from 1754 to 1829 appeared to Lamb "more homogeneous and trustworthy than that of most official observatories at the time" (Lamb and Johnson 1959: 132). Slowly, and with painstaking effort, Lamb began to coax standardized air pressure observations and measurements out of this plethora of records and, together with an assistant, to enter the resulting data on maps. While reliable barometers dated back to the eighteenth century, Lamb needed to develop and apply corrections for the temperature of unheated rooms, altitude above sea level, and archaic and inconsistent measuring units (Ogilvie 2010). By 1958, nearly two years into the project, Lamb realized his progress was too slow, and he applied to the Meteorological Office for further assistance. Sutton granted him the help of a young forecaster, Arnold Johnson, as well as a draughtsman, and between them they managed to complete the maps in time. In the end, Lamb's essay was awarded an honorable mention in the Napier Shaw Memorial Prize competition. First place was carried by S. K. Runcorn, Professor of Physics at Newcastle University, for a study of the magnetization of the rocks in the context of climate (Runcorn 1961). After more than two years of work, Lamb was somewhat comforted that the prize committee recognized his essay as "likely to afford considerable stimulus to the development of the subject" (Lamb 1997: 182).

Lamb's work between mid-1956 and 1959 focused on the construction of atmospheric pressure maps for months representing the extreme seasons (that is, January and July), covering as much of the globe as possible as far back in time as possible.[7] With the wealth of records available in the Meteorological Office archives, he was able to extend his maps back to 1760, making use for the first time of old records to reveal atmospheric circulation patterns of centuries ago. Together, the twenty maps (one for each decade from the 1760s to the 1950s) provided what Lamb and Johnson described as "a unified view of the known changes of temperature, extent of ice, ocean currents, rainfall and other phenomena" (Lamb and Johnson 1959: 132). By demonstrating global changes in the intensity and pattern of the general circulation of the atmosphere, they continued, their maps "disclos[ed] some of the essential laws of long-term behavior of the general circulation which have so far remained hidden" (Lamb and Johnson 1959: 104).

Primary among their findings was a correspondence between variations in the intensity of the general circulation and climate: weak periods in the general circulation corresponded to cooler climates (such as during the Little Ice Age from the 1760s until the 1830s, and the cooling after 1940) and strong periods in the general circulation corresponded to warmer climates (such as the warming in the early twentieth century, which had been pivotal to the shift away from a static view of climate). Their data collation also made it possible to test some of the many theories about atmospheric circulation and climatic change, including relationships between circulation and sea temperature, ice distribution, and volcanic activity (Lamb 1961: 132). Ultimately, Lamb and Johnson argued, such long time-scale and large region-scale studies could give insight into the anatomy of past climates and climatic changes. "I think this permits some optimism about the prospect of simplifying the welter of heterogeneous data on climatic changes and gaining better understanding," emphasized Lamb (Lamb 1959: 314). But, more importantly for the purposes of this chapter, this essay marks the beginning of Lamb's use of and appeal to historical climate data at a time when the research culture of climatology was in flux.

Lamb's philosophical approach to climate

With the rise of a dynamic conception of climate and the consequent rejection of climatic normal periods, the role, value, and legitimacy of historical climate records began to be questioned. If the past was not a reliable guide to the future, what role could these records serve, if any? Did they still have any explanatory or predictive value—or were they relics from an older era, of interest for understanding past events but with little or no application to the present or future? And with the emergence of computer models of the atmospheric circulation, where did (and where "should") historical climate records fit into the Meteorological Office's research programs?

For Lamb, the emergence of climate as a dynamic concept reinforced rather than undermined the need for historically-based investigations of climatic phenomena. Whilst Lamb agreed that past tables of climatic data could no longer be treated straightforwardly as reliable guides to the future, he still saw value in the information contained in these records.

> Climatology has a great need of the fullest possible, firmly established past historical record ... Without a record of climate's past behavior extending back over several repeats of the long-term processes of climatic variation, the subject would be in the situation of a branch of physics in which the basic laboratory observations of the phenomena to be explained had not been made. There can be no sound theory without such an observation record.
>
> (Lamb 1986: 17)

The fundamental questions facing the discipline, he emphasized again and again, necessitated careful reconstructions of the past record of climate—an observational base he thought integral to the scientific study of climate.

For Lamb, careful studies of the past climatic record were essential for developing numerical estimates of past climatic variations—data which could then "be explored for the workings of known perturbations and identifiable processes" (Lamb 1959: 303). This understanding of past climates would, Lamb hoped, allow for the identification of physical processes involved in climatic change and estimates of the magnitudes of climatic consequences due to diverse parameters and influences. Together, he argued, this knowledge was necessary "if we are to diagnose the symptoms of change now and some day, perhaps, be able to predict its course and amount" (Lamb 1964: 171). Using sources as varied as ships' logs, monastic chronicles, depth measurements from Viking burial grounds, and seasonal accounts from English vineyards, Lamb's research program pushed back into the past to reconstruct climatic data over centuries and even millennia. This work put as much emphasis on pre-instrumental data (e.g. diaries, crop information, sagas, and physical proxies) as on instrumental data (where records trace back to the development of the barometer in the seventieth century).[8] Lamb's use of historical climate records to study climatic change set him apart from the direction of the Meteorological Office from the mid-1960s on; that is, the burgeoning use of numerical models to represent and understand weather and climate patterns.

Lamb's commitment to historical climatology forms one part of his three-part philosophical approach to climate, developed through the 1950s and 1960s. Lamb was also committed to an interdisciplinary investigative methodology—one which pulled together experts from a wide variety of fields, natural as well as social sciences, humanities as well as the applied sciences, experts as well as amateurs, to provide a robust understanding of climate. This leads to a second set of questions that emerged in the post-war decades: With the emergence of a dynamic conception of climate, what was to be the relation of climate science to other fields of study? Where did the discipline's boundaries fall? Who could legitimately speak about climate and offer expert opinions, and who not? And how should the Meteorological Office's climate branches define and delineate themselves?

Lamb saw climatology as a comprehensive, manifold discipline pulling together "most aspects of the human environment" to shed light on climatic behavior and processes (Lamb 1969a: 1210). He regularly met and corresponded with art historians, glaciologists, botanists, historians, textual philologists, archaeologists and dendrochronologists, as well as undertaking field trips to locations including the Faroe Islands to uncover evidence pertaining to past climates. Lamb was also a voracious collector of newspaper clippings, church engravings, early modern almanacs, manorial accounts and landscape artwork—anything, in short, that could provide a direct or indirect window into past climatic conditions (Lamb n.d.2, "Old Weather Records").

Within this vision of climatology, numerical modeling of climate was notably absent. As numerical models gained more and more attention and resources at the Meteorological Office in the mid-to-late 1960s, Lamb began to voice trepidations about the role of computers in climate research. At a 1968 conference in Cambridge, England, he described computer models as "too unrealistic, as yet, for further use [since] much testing of the apparent validity of the results is needed," and warned that "we must be very wary for a long time yet in building elaborate constructions upon a [numerical model] base" (Lamb 1968b: 58). Following the appointment of a new director, Basil John Mason, at the Meteorological Office in 1965, the Office's climate modeling section expanded and the Office's research culture grew increasingly distant from Lamb's own priorities—to such an extent that in a 1973 commentary in *Nature* he lamented that many of his colleagues were convinced that climatic problems "will only be solved by mathematical modeling techniques [...] and that there is little point in attempting to reconstruct the past record of climate in any detail" (Lamb 1973: 397).

All three aspects of Lamb's philosophical approach to climate—his commitment to historical investigations, his desire for an interdisciplinary research culture, and his cautious stance towards numerical models—coalesced in his view of climatic prediction. At the Meteorological Office, too, climate prediction was a key issue under Mason's leadership—but not without debate. Which methods of climate prediction were to be trusted and imbued with authority, and which were not? What legitimate uses were there for climate prediction, and how authoritatively could the Office speak on the matter? How could epistemological bases for prediction be justified? And what culture of prediction was to guide the Office's work over the following years and decades?

As an organization with a strong service mandate, the Meteorological Office regularly answered inquiries about weather and climate from a host of sources: government, the military, agricultural interests, domestic and overseas industries, and the public, among others. In 1965, the year Mason took over as director, the Office answered over a million weather and climate inquiries from industry and from the public and fulfilled nearly 10,000 requests for climate data (Mason 1966: 383). To take but one example, Binnie & Partners, a British firm contracted to design Pakistan's Mangla Dam—today, the ninth largest dam in the world—appealed to the Meteorological Office to estimate future precipitation in the dam's catchment area and to advise on the diversion of the river during the monsoon season. And as severe weather events such as the hurricane force winds that pummeled Sheffield in 1962, damaging more than two-thirds of the city's houses, caught the attention of planners in government and industry, these inquiries increasingly focused on climate prediction and the potential effects of climatic change.

Climatology is facing problems that were not dreamt of a generation ago: no less than a demand—natural enough in an age of long-range forward planning and agriculture, industry, trade and government—for forecasts of the probable climatic trend over the next few years or even decades ahead, [wrote Lamb at the end of the 1960s:] the demand began to be felt as soon as it became widely known that significant climatic shifts had been taking place within the present century.

(Lamb 1969a: 1209)

For Lamb, the rapid advance of computer modeling combined with the growing demand for climatic predictions threatened to send climate research on a dangerous course. "There is an obvious economic call for forecasts of the climatic tendency over the decades ahead," he agreed, but "this need cannot be met until a scientific basis for such forecasts has been created"—and that basis, he continued, "must be the fullest and most specific possible knowledge of past climatic behavior" (Lamb 1964: 170). Lamb was particularly concerned that numerical modelers would jump the gun by using newly-developed models to provide long-term climatic predictions which might then be acted upon preemptively on the political scene.

The computer models of atmospheric behavior and other climatic areas may be unrealistic, and may therefore proceed too far and too fast on faulty basic assumptions, [he emphasized in *Nature*:] Such developments should be preceded by acquiring fuller and firmer factual knowledge.

(Lamb 1969a: 1215)

Boundary work at the Meteorological Office

By the mid-to-late 1960s, Lamb's cautious, and even critical, stance towards numerical climate models was at odds with the culture at the Meteorological Office. Under Mason's directorship, the Office invested heavily in computer modeling, namely numerical weather forecasting and, soon thereafter, numerical modeling of climate.

Soon after beginning his tenure as director on 1 October 1965, Mason decided to move numerical weather prediction beyond the research stage and make it operational—a decision that went against the advice of his senior scientists. A month later, on Monday, 2 November 1965, the Meteorological Office began issuing twice-daily numerical forecasts. Mason announced this breakthrough to the public at the Office's first press conference, where the assembled journalists and photographers were invited to watch the UK's first electronic routine weather chart slowly come off the printer and to take a souvenir copy of the chart home with them (Press Conference 1966: 28).

To Mason, a cloud physicist by training, computers offered a means of quelling criticism over the Meteorological Office's abilities and of laying out the organization's future according to his blueprint. Under pressure to improve forecasting, Mason saw computers as an objective tool for a traditionally subjective profession.

In 1966, the Office's forecasting ability was loudly criticized by, among others, members of the House of Commons and the House of Lords, *The Times* newspaper, and the International Federation of Airline Pilots Associations. "What steps [are] the Government [...] going to take to correct the deterioration in the standard of forecasts issued by the Meteorological Office?" demanded Lord Erroll of Hale in the House of Lords that summer (Sunnier Outlook for Weather Forecasts, 2 August 1966). In response, Mason declared that "the traditional, empirical and largely subjective methods of weather forecasting" would be replaced by "objective mathematical predictions made with the help of powerful electronic digital computers" (Mason 1978b: 297). Within a decade of his appointment, as Mason wrote in the *Meteorological Magazine*, "the Weather Service ha[d] been extensively modernized and re-organized around the very powerful IBM 360/195 computer system [...], the forecasting now being firmly based on the most advanced numerical models in operational use" (Mason 1978a: 129). Mason treated climate similarly.

> By far the most promising approach to understanding the prediction of climatic change lies in the construction of physico-mathematical models of the global circulation of the atmosphere and oceans treated as a combined geophysical system.
>
> (Mason 1976: 53)

This statement was supported in practice by the growing group of scientists working on numerical climate modeling at the Meteorological Office.

As the Meteorological Office concentrated climate resources on modeling, Lamb's research program received less and less support, both morally and financially. In 1963, two years before Mason's arrival, Lamb had won a promotion to senior principal scientific officer, freeing him from administrative duties and allowing him to concentrate fully on his research. But with Mason's new role as director, the Meteorological Office's culture changed and Lamb found it increasingly difficult to pursue his research agenda (see also Lamb 1966a). Invited to visit Hal Fritts' Tree Ring Research Laboratory at the University of Arizona and examine data on tree growth and climate stretching back more than 500 years, for example, Lamb applied to the Meteorological Office for travel funding, but was denied— an incident which, to Lamb, "proves in a glaring way just how the Meteorological Office denies support to [my] research" (Lamb, 1970). With his eye firmly on a computer-based research culture, Mason had little time for Lamb's research priorities, taking the position that

> the study of climatic change was a matter of developing numerical models on rather larger computers than those yet available and that the analysis of past events was irrelevant to this exercise, or at very best, extremely dull.
>
> (Clayton 1970)

Mason's language on these matters—language teeming with references to subjectivity and objectivity—is important for understanding his thinking, and likewise the attitudes he instilled at the Meteorological Office early in his tenure.[9] For Mason, the role of objectively-implemented physical theory in meteorology and climate research was critical to the future of those disciplines: numerical methods, he argued "are objective, logical, mathematical exercises based on a firm structure of physical theory", whilst traditional weather forecasting and historically-based climate work "depend heavily on the experience, skill and judgement of the individual human" (Mason 1973: 63). By the mid-to-late 1960s, Mason continued, the time was ripe to move away from the latter and to devote energy and resources to numerical techniques.

For Lamb, however, 'subjective' knowledge was integral to his scientific identity. He found knowledge that relied on the vast experience of individuals, who had devoted years or decades to studying and understanding past weather and climate patterns, and amassing and interweaving evidence from a wide variety of branches of knowledge, highly valuable. This was, he insisted, knowledge which could not be captured by strict computer processes. But at the Meteorological Office, support for this type of knowledge declined under Mason, who instilled what Lamb described as "unhelpful attitudes" towards historical and cultural ways of knowing (Lamb 1997: 200). "Few physical scientists [are] willing to read the literature, particularly that involving historical documentary accounts of details carefully recorded at the time, about past climate", continued Lamb: "History", he concluded, "is not their business" (Lamb 1997: 200).

Lamb vehemently opposed the way in which climate was subsumed by a 'hard' scientific approach under Mason's early tenure: Mason's push for modeling, Lamb thought, reduced climate to a purely physical phenomenon with no associations to culture, geography, or social or human history—a reductionist approach he found misguided at best and harmful at worst. A purely physical, or numerical, interpretation of climate, Lamb worried, negated other ways of knowing—ways which might prove crucial to a full understanding of climate.

With resources increasingly directed to numerical modeling and a focus on climate as a purely physical phenomenon, Lamb's interdisciplinary approach to climate also fell outside the boundaries demarcating the Meteorological Office's new culture. The Office's leadership, Lamb wrote to the University of East Anglia's Donald V. Osborne in 1966, suffered from "blind spots", from "a single-minded concentration on the mathematician's aspects—the dynamics of the atmospheric circulation and numerical forecasting" (Lamb 1966b). And, he continued in a pessimistic letter to Osborne's colleague Clayton three years later, "it does not seem likely that the state weather service can ever contemplate covering the interdisciplinary aspects adequately" (Lamb 1969b).

Lamb soon concluded that the Meteorological Office could no longer provide the support structures he needed to pursue his research. In 1971, with Clayton's support, Lamb left the Meteorological Office after 36 years of service and founded the Climatic Research Unit at the University of East Anglia—a unit dedicated to interdisciplinary and historically-based investigations of climate. By doing so, Lamb staked his reputation and success on a particular vision of climate research. The establishment of the Climatic Research Unit, and Lamb's efforts to win funding and support for the new unit from government, industry, and foundations, provide another window for understanding boundary work in British climatology at the time—but that, as they say, is another story.[10]

This chapter has shown how the emergence of distinctive cultures of prediction in UK meteorology and climatology was characterized by profound disagreements over the epistemic standards and practices by which climate and its changes could be comprehended. It has emphasized how, even two decades into the post-war period, the primacy of numerical computation was not taken for granted by all those who were professionally interested in long-term climatic change. As such, this history points to the significance of institutionalization and institutional cultures in understanding the character of cultures of prediction; as an increasingly marginalized figure in the Meteorological Office, Lamb serves as a reminder that the institutionalization of numerical prediction often occurred to the cost of other forms of scientific practice, while also pointing to the dynamic roles played by key individuals, like Lamb and Mason, in shaping the institutional cultures which allowed particular cultures of prediction to flourish. The epistemic and cultural authority of computational prediction was far from pre-determined; it was rather the product of significant instances of boundary work and disagreement around the proper practices of comprehending a dynamic climate.

Notes

1   Here, Cohen et al. refer specifically to the hegemony of general circulations models.
2   This stable picture of climate was punctured only by known radical variations in the past, such as the ice ages and the warm epoch of the Middle Ages. However, climatology was little affected by these debates, which were studied primarily by geologists and glaciologists.
3   Recent literature identifies this warming period as beginning in the 1910s or 1920s (Brönnimann 2009).
4   See, for example, "Retreat of the Cold," 29 October 1951; "Ice-free Arctic?," 17 May 1954.
5   The 1950s and 1960s were also a period of upheaval in the Meteorological Office for other, but no less important, reasons: under director Sir Oliver Graham Sutton, the Office underwent a major re-organization of its research and service divisions as well as unification under a single new roof in Bracknell. In these decades, too, numerical weather forecasting played an increasingly important role at the Office, as did rising public, political and industrial inquiries about weather- and climate-related topics (Hall 2012; Walker 2012).
6   For a biographical overview of Lamb, see Martin-Nielsen (forthcoming).
7   By the time they submitted their essay, Lamb and Johnson had completed maps for every January since 1760 and were nearing completion of a corresponding series of July maps.
8   Lamb repeatedly emphasized the difficulties of working with such data (Lamb 1959; Lamb 1968a).
9   Here, an "objective" method of forecasting is "one which depends only on the initial data and will produce the same answer whoever prepares it; the method will not call

for any judgement on the part of the forecaster" (Meteorological Office Discussion 1959: 207). Note the importance of the forecaster's judgement being removed from the forecast—this was precisely Mason's aim.

10 This story is being pursued under the framework of the project "*Shaping Cultures of Prediction: Knowledge, Authority, and the Construction of Climate Change*" at Aarhus University's Center for Science Studies.

Acknowledgments

The author wishes to thanks the archivists and librarians who went out of their way to assist with the research for this chapter: Bridget Gillies (University of East Anglia Archives), Andrew Watt, Mark Beswick and Joan Self (National Meteorological Library and Archive), and Alan Ovenden (Climatic Research Unit). Many thanks also to the organizers and participants of the "UK Climatology 1960–1985 and the Emergence of Climate Modelling" workshop at King's College London in January 2015 and to Matthias Heymann for reading many drafts of this paper.

References

Brönnimann, S. (2009) "Early Twentieth-Century Warming," *Nature Geoscience* 2, pp. 735–736.

Clayton, K. M. (1970) Letter to S. Zuckerman, 11 June 1970 (UEA Archives, J. R. Jones 1966–1982 Collection, Climatic Research Unit 1963–1971 Subcollection, Document UEA.Jones.40.131).

Cohen, S., Demeritt, D., Robinson, J. and D. Rothman (1998) "Climate Change and Sustainable Development: Towards Dialogue," *Global Environmental Change* 8, pp. 341–371.

Davies, T. (2004) "Hubert Horace Lamb (1913–1997)," *Oxford Dictionary of National Biography*, Oxford, UK: Oxford University Press.

Doel, R. E. (2009) "Quelle Place Pour les Sciences de l'Environnement Physique dans l'Histoire Environnementale?" *Revue d'Histoire Moderne et Contemporaine* 56, pp. 137–164.

Edwards, P. N. (2011) "History of Climate Modeling," *Wiley Interdisciplinary Reviews: Climate Change* 2, pp. 128–139.

Gieryn, T. F. (1983) "Boundary-work and the Demarcation of Science from Non-Science: Strains and Interests in Professional Ideologies of Scientists," *American Sociological Review* 48, pp. 781–795.

Gieryn, T. F. (1999) *Cultural Boundaries of Science: Credibility on the Line*, Chicago, IL: University of Chicago Press.

Guillemot, H. (2007) "Les Modèles Numériques de Climat," A. Dahan (ed.) *Les Modèles du Futur: Changement Climatique et Scénarios Economiques: Enjeux Politiques et Economiques*, Paris, France: La Decouverte.

Hall, A. (2012) *Risk, Blame and Expertise: The Meteorological Office and Extreme Weather in Post-War Britain*, (PhD Thesis), Manchester, UK: University of Manchester.

Heymann, M. (2009) "Klimakonstruktionen: von der klassichen Klimatologie zur Klimaforschung," *NTM Journal of History of Sciences, Technology and Medicine* 17, pp. 171–197.

Heymann, M. (2010) "The Evolution of Climate Ideas and Knowledge," *Wiley Interdisciplinary Reviews: Climate Change* 1, pp. 581–597.

Heymann, M. and J. Martin-Nielsen (2013) *Shaping Cultures of Prediction: Knowledge, Authority, and the Construction of Climate Change*, (Proposal to the Danish Council for Independent Research – Humanities), Arhus, Denmark: University of Arhus.

Jones, P.D. (1997) "Obituary: Hubert Horace Lamb," *Quarterly Journal of the Royal Meteorological Society* 123, pp. 2165–2166.

Kaempffert, W. (1952) "Oldtimers May Be Right When They Tell Us That the Climate is Getting Warmer," *The New York Times*, 30 November 1952.

Kington, J. A. (2008) "Hubert H. Lamb: A Review of His Life and Work," *Weather* 63, pp. 187–189.

Lamb, H. H. (1959) "Our Changing Climate, Past and Present," *Weather* 14, pp. 299–318.

Lamb, H. H. (1961) "Climatic Change Within Historical Time as Seen in Circulation Maps and Diagrams, 1961" (Classmark P216—LAMB, National Meteorological Library and Archive, Exeter, England).

Lamb, H. H. (1964) "Britain's Climate in the Past: A Lecture Given to Section X of the British Association for the Advancement of Science at Southampton on September 2, 1964," (reprinted in Lamb, H. H. (1964) *The Changing Climate: Selected Papers*, London: Methuen & Co.).

Lamb, H. H. (1966a) Letter to The Secretary to the University of East Anglia, 5 October 1966 (UEA Archives, J. R. Jones 1966–1982 Collection, Climatic Research Unit 1963–1971 Subcollection, Document UEA.Jones.40.6).

Lamb, H. H. (1966b) Letter to D.V. Osborne, School of Mathematics and Physics, UEA, 29 December 1966 (UEA Archives, J.R. Jones 1966–1982 Collection, Climatic Research Unit 1963–1971 Subcollection, Document UEA.Jones.40.3).

Lamb, H. H. (1968a) "Ancient Units Used by the Pioneers of Meteorological Measurements," *Weather* 41, pp. 230–233.

Lamb H. H. (1968b) "Investigation of the Climatic Sequence: A Meteorological-Empirical Approach – Lecture Given at the SCAR Conference on Quaternary Studies in the Antarctic, Cambridge, England, July 1968" (National Meteorological Library & Archive, Pamphlet P327(LAMB), Item ID 270707-1001).

Lamb, H. H. (1969a) "The New Look of Climatology," *Nature* 223, pp. 1209–1215.

Lamb, H. H. (1969b) Letter to K. M. Clayton, Dean of the School of Environmental Studies, UEA, 26 March 1969 (UEA Archives, J. R. Jones 1966–1982 Collection, Climatic Research Unit 1963–1971 Subcollection, Document UEA.Jones.40.1).

Lamb, H. H. (1970) Letter to K. M. Clayton, School of Environmental Studies, UEA, 10 April 1970 (UEA Archives, J. R. Jones 1966–1982 Collection, Climatic Research Unit 1963–1971 Subcollection, Document UEA.Jones.40.86).

Lamb, H. H. (1973) "Whither Climate Now?" *Nature* 244, pp. 395–397.

Lamb, H. H. (1986) "The History of Climatology and the Effects of Climatic Variations on Human History," *Weather* 41, pp. 16–20.

Lamb, H. H. (1988) *Weather, Climate and Human Affairs: A Book of Essays and Other Papers*, London and New York: Routledge.

Lamb, H. H. (1997) *Through All the Changing Seasons of Life: A Meteorologist's Tale*, Norfolk, UK: Taverner Publications.

Lamb, H. H. (n.d.1) "Mapping Historical Weather and Past Climates" (Box 3—Climate History, H. H. Lamb Archive, University of East Anglia Archives, Norwich, England).

Lamb, H. H. (n.d.2) "Old Weather Records of Known Date (New Style) and Historic Interest Dates, From the 8th-20th Centuries, Notebook/Diaries" (Box 6—Diaries, data sources and bibliographies, H. H. Lamb Archive, University of East Anglia Archives, Norwich, England).

Lamb, H. H. and A. Johnson (1959) "Climatic Variation and Observed Changes in the General Circulation, Parts I and II," *Geografika Annaler* 41, pp. 94–134.

Lamb, H. H. and A. Johnson (1960) "The Use of Monthly Mean 'Climate' Charts for the Study of Large Scale Weather Patterns and Their Seasonal Development," *Weather* 15, pp. 83–91.

Lamb, H. H. and A. Johnson (1961) "Climatic Variation and Observed Changes in the General Circulation, Part III," *Geografika Annaler* 43, pp. 363–400.

Martin-Nielsen, J. (2015) "Ways of Knowing Climate: Hubert H. Lamb and Climate Research in the UK," *Wiley Interdisciplinary Reviews Climate Change* 6, pp. 465–477.

Mason, B. J. (1966) "The Role of Meteorology in the National Economy (Based on a Lecture Given at the Summer Meeting of the Royal Meteorological Society in Brighton, 26 July 1966)," *Weather*, 21, pp. 382–393.

Mason, B. J. (1973) "The Application of Computers to Weather Forecasting," *Physics in Technology* 4, pp. 63.

Mason, B. J. (1976) "The Nature and Prediction of Climatic Changes," *Endeavour* 35, pp. 51–57.

Mason, B. J. (1978a) "The Future Development of the Meteorological Office," *The Meteorological Magazine* 107, pp. 129–140.

Mason, B. J. (1978b) "Recent Advances in the Numerical Prediction of Weather and Climate, A Lecture Delivered on 27 April 1978," *Proceedings of the Royal Society of London* 363, pp. 297–333.

Ogilvie, A. E. J. (2010) "Historical Climatology, Climatic Change, and Implications for Climate Science in the Twenty-First Century," *Climatic Change* 100, pp. 33–47.

Runcorn, S. K. (1961) "Climatic Change Through Geological Time in the Light of the Palaeomagnetic Evidence for Polar Wandering and Continental Drift," *Quarterly Journal of the Royal Meteorological Society* 87, pp. 282–313.

Sutcliffe, R. C. (1955) "The Meteorological Office Faces the Future: Scientific Research and Development," *Meteorological Magazine* 84, pp. 183–187.

Walker, M. (2012) *History of the Meteorological Office*, Cambridge, MA: Cambridge University Press.

Weart, S. R. (1998) "Climate Change, Since 1940," in Gregory Good (ed.), *Sciences of the Earth: An Encyclopedia of Events, People, and Phenomena* 1, pp. 99–104.

Weart, S. R. (2010) "The Development of General Circulation Models of Climate," *Studies in History and Philosophy of Modern Physics* 41, pp. 208–217.

Sources with no identified author

"Retreat of the Cold," *Time*, 29 October 1951.

"Ice-free Arctic?'," *Time*, 17 May 1954.

"The Napier Shaw Memorial Prize: Announcement of a Second Competition," (1956) *Quarterly Journal of the Royal Meteorological Society* 82, p. 383.

"Meteorological Office Discussion, 16 March 1959: Objective Methods of Local Forecasting," (1959) *The Meteorological Magazine* 88, p. 207.

"Sunnier Outlook for Weather Forecasts," *The Times* (London), 2 August 1966, p. 10.

"Press Conference" (1966) *Meteorological Magazine* 95, pp. 28–30.

Hansard, House of Lords Debate, 30 November 1978, Vol. 396, cc. 1442–1470, para. 1468.

6 From heuristic to predictive

Making climate models into political instruments

Matthias Heymann and Nils Randlev Hundebøl

Introduction

In recent years, climate modeling has emerged as the leading method of climate research and the predominant approach for producing predictive knowledge about climate. Climate modeling is widely identified as enjoying hegemonic status in present-day climate knowledge production and use (e.g. Shackley et al. 1998). Underlying the dominance of climate modeling and its uses in the production of climate knowledge are fundamental decisions about which types of knowledge are important, which epistemic standards are used to judge that knowledge, and which applications of that knowledge are regarded as useful and socially relevant. This paper aims to illuminate a critical juncture in the historical development of climate modeling: the shift from heuristic modeling to predictive climate modeling.

Climate models initially served heuristic purposes to investigate and better understand the processes governing climate and its variations. In the 1970s, a new generation of climate modelers pushed the development of climate models for the long-term prediction of global warming.

The character of the climate science that became increasingly visible around 1970 involved a number of significant and far-reaching reorientations, which in sum represented a fundamental cultural shift: the perception of environmental problems gained importance and problem perception turned from regional and national to global; a number of scientists recognized global climate change as a key concern, and the language of key climate scientists became a language of concern; climate modeling soon received primary attention for the production of predictive knowledge on the climate; and climate science in general and climate modeling in particular became increasingly politicized.

Hence, this article reviews a decisive moment in the emergence of climate modeling as a culture of prediction. It shows how social interests were absorbed and mobilized to establish climate prediction as a core interest in climate modeling. These efforts required an adjustment of models and modeling practices to facilitate the application of climate models for the scope of prediction. Furthermore, as climate models did not represent realistic representations of the atmosphere and were deemed uncertain, their application for predictive purposes required effective ways of domesticating uncertainty to convince scientific, political and public audiences about the reliability and usefulness of the predictive knowledge to be produced.

Authors such as Paul Edwards and Spencer Weart have described the emergence of global climate change as a political issue, the politicization of climate science and the development of climate modeling as a policy tool. The dominant narrative emphasizes the role of influential individual climate scientists, who "helped establish simulation modeling as a legitimate source of policy-relevant knowledge" (Edwards 2010: 359). Science-based and policy-oriented reports such as the Report of the US Presidential Scientific Advisory Committee of 1965 (PSAC 1965), the Study of Critical Environmental Problems (SCEP 1970) and the Study of Man's Impact on Climate (SMIC 1971), which were written by scientists, played a fundamental role. These reports, "published as elegant trade books, bound between sleek black covers, rather than as academic volumes" (Edwards 2010: 361) demonstrated attention to possible effects of an increase of atmospheric carbon dioxide (CO_2) levels and reflected and served the broader cultural trend of concern with regard to global environmental problems caused by humans.

PSAC recommended extended monitoring of CO_2 and global temperature as well as further work on climate models. The SCEP report explicitly declared that the risk of global warming was "so serious that much more must be learned about future trends of climate change" (SCEP 1970: 12). Both the SCEP and the SMIC "presented GCMs [General Circulation Models] and other models as the principal sources of climate knowledge" (Edwards 2010: 361). Earth Day in 1970 and the United Nations Conference on the Human Environment (UNCHE) in Stockholm in 1972 powerfully marked and publicized an era of emerging environmentalism and, as a part of it, propagated the need to better understand "the causes of climate change and whether these causes are natural or the result of man's activities" (UNCHE 1973: 21, recommendation 79). As a result, GCMs were put into a new arena and attained a new mission and meaning by becoming policy tools.

Heuristic and predictive modeling

Historian of science, Amy Dahan Dalmedico, has distinguished between heuristic and predictive modeling for the case of numerical weather simulation (Dahan 2001). She argued that modeling efforts for understanding atmospheric processes must be distinguished from the aim of calculating useful predictions. These different types of modeling not only entailed different applications of models and different uses of modeling results, they involved different priorities and research tasks as well as different practices and strategies.

The production of predictive knowledge requires a pooling of resources to address problems defined by this ultimate goal. Theory needs to be developed and adjusted for the specific scope of prediction, data need to be collected and processed accordingly, problems need to be prioritized and those problems estimated less fundamental for meaningful prediction relegated to later treatment. The distinction between heuristic and predictive climate modeling is not always sharp. The aim of predictive modeling requires heuristic forms of research in the development of appropriate theory and models. And, like all physical theory, heuristic modeling involves the calculation of specific predictions to be compared to observations to deepen theoretical understanding.

Still, the decision to focus on the improvement of theoretical understanding (and hence the improvement of model representations) or on the use of models for the production of useful predictions represents a decisive juncture. Numerical weather forecasting is a telling case. When the treatment of meteorological problems with the help of computers was discussed in the mid-1940s, project supporters had different goals. Francis W. Reichelderfer, head of the US Weather Bureau, and military meteorologists were interested in forecasting. Carl-Gustav Rossby and other academic meteorologists aimed at pushing meteorological theory. Vladimir Zworykin, who first suggested the treatment of meteorology by computers, and John von Neumann "wanted a meteorological theory amenable to an attack by computer," an advance that would ultimately allow for forecasting and weather control (Harper 2008: 104). The Meteorology Project at Princeton started with the very explicit goal of accomplishing weather forecasting based on the laws of physics (Nebeker 1995: 137).

Jule Charney was the ideal person to push von Neumann's approach. Kristine C. Harper writes that for Charney "weather forecasting was primarily a computing problem that required 'one intelligent machine and a few mathematico-meteorological oilers'" (Charney 2008: 116). Charney worked as an oiler of atmospheric theory and developed a "filtering scheme," a radical approximation and simplification of the differential equations to make them easier to solve by numerical integration. He ingeniously simplified the problem in a target-oriented way, a strategy that eventually facilitated the generation of a meaningful weather prediction after only two years of work. Von Neumann and Charney had a keen interest in theory, but focused on theory of a type that would work for useful applications. Their work aimed at and focused on predictive modeling, with heuristic modeling for better understanding pursued only in so far as it served this goal.

While Jule Charney (like von Neumann) thought predominantly in terms of solving the "forecasting problem" and retained a belief and interest in long-term forecasting (von Neumann pursued the idea of an "infinite forecast"), he also developed the philosophy of heuristic modeling and applied it to his work. Like Edward Lorenz, he promoted the idea of experimenting with models and deliberately accepting even simplifications so radical that a model could hardly be considered a representation of the atmosphere. Charney suggested a hierarchy of models in which individual models could serve to test the understanding of only specific processes. The model space served as a laboratory for experimentation, and the models resembled an idealized experimental apparatus in the laboratory. Running experiments with this apparatus allowed valuable insights, informed theoretical reasoning and improved scientific understanding.

Climate modeling as heuristic science

It was such an experiment that meteorologist Norman Phillips, one of Charney's associates in the Meteorology Project, undertook in 1955. Phillips wanted to test whether a weather forecasting model could be used to simulate atmospheric conditions over a period of time longer than a few days. The model he used and the initial conditions he defined were far from realistic, and it was unclear whether results would turn out to be meaningful or not. They did. His model simulation produced patterns of atmospheric circulation that resembled observed patterns. Phillips was aware of the crude simplifications in his experiments, but nevertheless concluded that "the verisimilitude of the forecast flow patterns [with observed patterns] suggest quite strongly that it contains a fair element of truth" (quoted in Lewis 1998: 51). This type of atmospheric model was soon called a GCM.

The results provoked enormous excitement. Yale Mintz, leader of one of the first climate modeling groups at the University of California, Los Angeles, explained in 1958:

> Although there are details that are wrong, the overall remarkable success achieved by Phillips in using the hydrodynamical equations to predict the mean zonal wind and mean meridional circulations of the atmosphere must be considered one of the landmarks of meteorology.
>
> (Mintz quoted in Arakawa 2000: 8)

Phillips's experiment not only showed that Vilhelm Bjerknes's so-called primitive equations could serve as a basis for regional weather forecasting, but also helped to explain major features of the atmospheric circulation. British meteorologist, Eric Eady, pointed out a second fundamental conclusion:

> Numerical integrations of the kind Dr. Phillips has carried out give us a unique opportunity to study large-scale meteorology as an experimental science.
>
> (Eady quoted in Lewis 2000: 117)

Charney concluded that the computer could be used as an inductive machine (Dahan 2001: 405). It served the exploration of physical processes and the production of knowledge about these processes by model experimentation, because material experimentation was impossible. Paul Edwards considered Phillips's experiment to be a "proof of concept," which encouraged a new generation of theoretical meteorologists to develop models and pursue simulation experiments to investigate and improve scientific understanding of the atmosphere (Edwards 2010: 152). The models developed "aimed not at forecasting but at characterizing how the atmosphere and the oceans process solar energy" (Edwards 2010: 143). Models served as useful instruments for basic research (Weart 2010: 210).

Von Neumann immediately attempted to capitalize on Phillips's experiment and concluded at a hastily organized conference that "we feel that we are now prepared to enter into the problem of forecasting the longer-period fluctuations of the general circulation" (Neumann quoted in Lewis 2000: 94). But rather than forecasting, it was the heuristic mode of modeling that characterized the first generation of developers of General Circulation Models. Phillips's experiment had opened up a new research direction. Complex systems like weather and climate had become accessible to quantitative understanding based on the laws of physics. It motivated a small scientific community of mainly theoretical meteorologists to tackle the problem piece by piece and contribute to what William Kellogg and Stephen Schneider later called a "theory of climate" (Kellogg and Schneider 1974).

The early history of climate modeling has been investigated in some detail (Edwards 2010, 2011; Weart 2010, 2014). Syukuro Manabe and his group at the Geophysical Fluid Dynamics Laboratory (GFDL) focused on the radiative-convective transfer of energy in the atmosphere and on atmosphere–ocean coupling. According to Edwards, GFDL took a "long-range view of the circulation modeling effort" (Edwards 2010: 154) to focus on an appropriate understanding of the atmosphere.

> Their research strategy used GCMs to diagnose what remained poorly understood or poorly modeled, and simpler models to both refine theoretical approaches and modeling techniques, in an iterative process (Edwards 2010: 154). Strict attention to developing physical theory and numerical methods before seeking verisimilitude became a hallmark of the GFDL modeling approach.
>
> (Edwards 2011: 132)

Akio Arakawa at the University of California in Los Angeles, hired by Yale Mintz, developed approaches to long-term numerical integration and contributed to the problem of representing the effect of clouds in general circulation models. In an interview several decades later, Arakawa emphasized that a "Modern Theory of the General Circulation of the Atmosphere" was his "big project" (Edwards 1997). But he was little interested in application. Further, but less influential early climate modeling efforts were pursued by Warren Washington and Akira Kasahara at the National Center for Atmospheric Research (NCAR) and by Cecil E. Leith at the Lawrence Livermore National Laboratory.

By 1964, Yale Mintz and Akio Arakawa had produced a global GCM with only two vertical layers, but a realistic geography. By 1965, Manabe's group had developed a three-dimensional global model with nine vertical levels. The model was still highly simplified, with no geography included, but able to provide quite realistic simulation results of processes such as transfer of heat and moisture. A Committee of the National Academy of Sciences concluded in 1966 that the best models simulated atmospheres with gross features "that have some resemblance to observation" (National Academy of Sciences, 1966, 65–67 quoted in Weart 2014).

Climate science meets politics

While von Neumann strategically pointed out the political value of his research efforts, the pioneers of climate modeling did not in any way articulate strong political interests related to their work. If they linked modeling of climatic processes to political issues, it served to justify their work and argue for funding rather than express political interest or concern. Typically, to take an example quoted by Spencer Weart, Yale Mintz provided a list of possible uses for his and Arakawa's computer model in 1965.

> Mintz showed an interest mainly in answering basic scientific questions. He also listed long-range forecasting and 'artificial climate control'—but not greenhouse effect warming or other possible causes of long-term climate change.
>
> (Weart 2014 referring to Mintz 1965: 153)

The CO_2 question was not of major interest in climate modeling between 1955 and 1970, and certainly not the driver of research in this field. The interest of climate modelers focused on understanding the climate system and developing a representation of the physical processes determining climatic phenomena. But this was to change—and the social and political contexts played a crucial role.

Joseph Smagorinsky recalled in an interview that the committee meetings for the PSAC report prompted him to ask Manabe to add CO_2 to his radiation model (Weart 2014: fn. 39). This statement also suggests that he (and Manabe) did not think much about the increase of CO_2 in the atmosphere before being alerted by a high-ranking policy committee. When Syukuro Manabe and Richard Wetherald performed the test and used their one-dimensional model to simulate what would happen if the level of CO_2 doubled, they came up with their famous result that global temperature would rise by roughly 2°C (Manabe and Weatherald 1967). This result, though preliminary, simulated with a model and not validated by observation, became an important resource in the emerging climate change discourse.

Wallace Broecker later recalled that it was this 1967 paper "that convinced me that this was a thing to worry about" (Weart 2014). The possibility of a warming climate prompted increasing concern among a number of atmospheric scientists. It was the preparations for the UNCHE, to be held in Stockholm in 1972, that motivated leading members of the global scientific community to take stock of the environmental situation and urge political action. An insistent and effective lobby was organized by Carroll Wilson of the Sloan School of Management at MIT, supported by William Matthews of MIT's Department of Civil Engineering, who in 1970 convened a month-long study session on global environmental problems, including the problem of global climate change. This session was attended by sixty-eight scientists along with numerous supporting consultants and observers. One year later, a similar three-week study session was held in Stockholm in the summer of 1971, which exclusively focused on "man's impact on climate." These sessions resulted in the famous SCEP (1970) and SMIC (1971) reports and an additional report of SCEP background papers (Matthews et al. 1971). All reports were rushed to publication only few months after the events to indicate political urgency and serve as preparatory documents for the UN Stockholm conference.

NCAR atmospheric division director, William Welch Kellogg, one of the organizers and editors of the volume of background papers for the SCEP study, concluded in his chapter that

> there is the haunting realization that man may be able to change the climate of the planet Earth. This, I believe, is one of the most important questions of our time, and it must certainly rank near the top of the priority list in atmospheric science.
>
> (Kellogg 1971: 123)

Kellogg's chapter was titled: "Predicting the Climate," thus illuminating what he considered the major task to be tackled. With this term "prediction" he had a specific meaning of the word in mind: the prediction of long-term climatic change. Climate models, he argued, had to serve not only scientific roles, but also the important social role of providing politically relevant knowledge about future climate change. Edward Lorenz was represented in the same report with a paper he had published one year before. He pointed to fundamental shortcomings in the existing models. Still he considered "the mathematical model of the atmosphere a new and powerful tool for studying the phenomenon of climatic change" and suggested "perhaps there should be a center for climatic-change hypothesis testing" (Lorenz 1971: 188).

Reports like the PSAC, SCEP and SMIC proved influential by both spreading a spirit and a language of political concern and raising interest in and attention to a certain direction of future research. Atmospheric scientists effectively claimed political responsibility and, consequently, demanded resources for the investigation of future global warming.

The problem of climate prediction

The goal Kellogg declared in 1970 (and published in 1971), the "prediction of climate," was far from easy to accomplish and controversial from the outset. Edward Lorenz's theoretical work questioned the possibility of climate prediction altogether. Lorenz had investigated the chaotic structure of the atmosphere, which caused indeterminacy and limited its predictability. This was the deeper reason for the fact that numerical weather prediction was not able to achieve useful prediction beyond a few days ahead. In his research on system predictability, Lorenz concluded that a mathematical system like a climate model could experience shifts between different stable states without a change in the boundary conditions of the system. This result compromised the very idea of prediction. Lorenz defined three different kinds of mathematical systems: the "transitive" system, where only one stable solution existed; the "intransitive" system, where several different stable solutions existed; and the "almost intransitive" system, where one main state existed, but over long periods of time the system could deviate far from that stable state (Lorenz 1968). Before a target system (such as climate) was fully described by a mathematical model and the mathematical representation tested, it was not possible to decide which of the three kinds of systems the target system represented.

Prediction is possible for transitive systems, for which the state of a mathematical model after a change in boundary conditions can be determined and a causal mechanism identified, that is a deterministic prediction can be made. For intransitive systems prediction is impossible, as the system itself produces irregular changes and perturbations even with stable boundary conditions. Such perturbations superpose and, thus, hide system changes caused by changes in the boundary conditions (such as a rise in CO_2 levels). Lorenz could not decide with any certainty which type of system the climate system was, but carefully concluded that "almost intransitivity becomes an attractive hypothesis" for mathematical models of a system like the climate, which comprises more processes than those of the atmosphere alone (Lorenz 1970: 328)—consigning the question of system predictability to limbo.

At a GARP (Global Atmospheric Research Program) study conference on climate science and modeling in Stockholm in 1974, Lorenz distinguished two kinds of climate predictions. Accurate simulation with a climate model from a given initial climatic state to a future climatic state would be a "climatic predictions of the first kind." This type would resemble weather prediction, except that the prediction would emphasize months and seasons rather than hours and days. In this case, the differences in climatic states could be traced back to and explained by differences in the initial atmospheric conditions. However, in 1957 Lorenz considered climate predictions of the first kind impossible, at least with the current state of climate models and climate science. Alternatively, he suggested "climatic predictions of the second kind." This approach treated climate-relevant processes like changing CO_2 concentrations and atmosphere-ocean interactions as external to the atmospheric system and represented as mere boundary conditions impacting on it. Predictions of the second kind would predict the effect of a change in the equilibrium climate of a model as a response to a change in boundary conditions of the model, for example, CO_2 levels (Lorenz 1975).

For both kinds of prediction Lorenz's approach would allow valid climate prediction provided that the target system climate, like the mathematical models describing it, was transitive. Investigations of changed global mean temperature due to increased CO_2 concentrations, as provided by Manabe and Wetherald (1967, 1975), was in Lorenz's terminology a prediction of the second kind, and thus valid only under the condition of climate system transitivity. Given the simple character of the mathematical models that produced stable equilibria of climate compared to the complex and ever fluctuating natural climate, Lorenz saw good reason to be cautious about climate model predictions, because the climate system appeared to be almost intransitive. Furthermore, predictions of the second kind were in Lorenz's eyes only a second choice: "we would desire a prediction of the first kind, but, lacking sufficient skill, we might have to settle for a prediction of the second kind" (Lorenz 1975: 132).

Kellogg was aware of the challenges that Lorenz's work posed to his call for climate prediction. In fact, in his report to the World Meteorological Organization (WMO) he stated that the problem of intransitivity and almost intransitivity "hangs like a black cloud over those of us who are seeking to throw some light on the study of human influences [on earth's climate]" (Kellogg 1997: 4). But Kellogg rejected the assumption of intransitivity. Instead of the "the pessimistic view of Lorenz" he favored a "more pragmatic view" as Cecil E. Leith had suggested, because "for sufficiently small changes about the present climate" a prediction of the second kind should be possible and "appropriate" (Leith 1975: 140).

Lorenz expected current models to be too stable, failing to reflect the variability of climate, in his words "almost intransitivity," which only more developed models would be able to represent. Kellogg, in contrast, trusted climate to be transitive because simulations of current models appeared stable and settled in defined equilibrium states, which in his eyes compared well enough with observational evidence. This model behavior clearly reflected a transitive system. Kellogg certainly knew that current models represented strong simplifications. He discussed the fact that current models did not account for major feedback loops such as the cloud-temperature-albedo loop and the atmosphere-ocean-sea surface temperature loop. But he argued that these loops would only result in weak damping effects and, therefore, not cause major changes compared to model predictions (Kellogg 1977: 4, 5).

The different assessments of Lorenz and Kellogg depended on the question of whether the climate system behaves too chaotically (Lorenz) or deterministically enough (Kellogg) to make deterministic prediction reliable. These assessments also resulted from different interests. Lorenz took a general interest in climatic variability on the annual to decadal scale, which included episodes of drought, crop failures, and famine. Kellogg, continuing the work in the SCEP and the SMIC, took an interest in human pollution of the environment and focused on global warming due to the emission of greenhouse gases as the principal problem on a decadal to centennial scale. If the equilibrium response to rising CO_2 levels in a transient climate model was larger than the natural variation, predictions should prove their merit; if only by suggesting a significant trend for the future climate rather than by providing precise numbers.

In his report, Kellogg published a graph showing such a prediction (see Figure 6.1), one of the first of its kind, which was meant "to present a probable scenario of the future so that the decision makers of the world can begin to formulate their various value judgments" concerning the societal implications of future climatic change (Kellogg 1977: 1); other long-term projections based

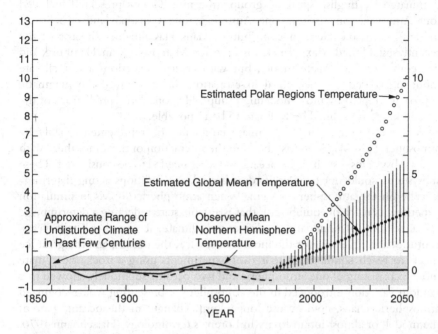

Figure 6.1 Estimates of past and future variations in global mean temperature.
Source: Kellogg 1977: 24.

on other types of evidence were published by Mitchell (1977) and Broecker (1975). He predicted a change in global mean temperatures by 3–4°C by the year 2050. This prediction was based on the results of climate models with additional modifications to account for feedbacks not accounted for in the models (see also Heymann 2012). Kellogg acknowledged that his analysis was simplistic, but still concluded that the effects of mankind

> will be considerably larger than the expected natural changes [...] This *should* be a useful piece of information [Kellogg's emphasis]. It may turn out that the extreme warming that could conceivably occur toward the latter part of the next century will be deemed "unacceptable" by the nations of the world and that strong international action will then be taken to drastically cut down the burning of fossil fuels or to institute counter-measures against the warming.
>
> (Kellogg 1977: 32, 33)

Both Kellogg's and Lorenz's interpretations were soon questioned. At the GARP Study Conference on Climate Models, Performance, Intercomparison, and Sensitivity Studies in Washington, DC in April 1978, climate modeling groups from across the world met to compare their climate models and the predictive performance of these models (Gates 1979). The conference was dominated by English-speaking groups (the nine GCMs presented included four American, one British, one Australian, one Canadian and two Soviet models). The sole German contributor, Klaus Hasselmann, director of the recently established Max Planck Institute for Meteorology in Hamburg, did not come with a climate model, but with critical reflections about climate modeling. Based on theoretical investigations he challenged the pragmatic approach adopted by most modeling groups. He considered predictions of the second kind, as outlined by Kellogg, to be impossible.

According to Hasselmann, climate could not be represented reliably by atmospheric GCMs (AGCMs), because the interaction of the atmosphere with other subsystems—such as the oceans and ice—needed to be considered. These subsystems could not be simplified as boundary conditions acting deterministically on the atmospheric system. When atmospheric GCMs in simulation experiments reached equilibrium, that is a stable state reflecting a deterministic causal relation of boundary condition and climate, it would still take years, centuries and millennia for the models of sea-ice, the oceans, and the great ice sheets to reach equilibria. Theoretical experiments using a stochastic climate model, consisting of one atmospheric and two oceanic submodels, showed that random variation introduced in the atmosphere, representing well-recognized atmospheric chaos, could cause longer-term changes in the oceanic part of the model, or almost intransitivity in Lorenz's terminology (Hasselmann 1976; Frankignoul and Hasselmann 1977; Lemke 1977). Subsystems like the oceans or sea ice could not be treated as stable boundary conditions, as had generally been the case.

Hasselmann concluded that "the interpretation of atmospheric GCM response in a climatic context requires the imbedding of the atmospheric model in longer timescale models" (Hasselmann 1979: 1048). Climate models needed to represent the interactive exchanges between climate subsystems before they could reliably address questions of climate prediction. This was not the case for the type of models discussed at the Washington conference and used by Kellogg for his prediction. Hasselmann's conclusion undermined the hope for transitivity of the climate system. Lorenz's suggestion of two types of deterministic prediction broke down. As a consequence, Hasselmann suggested a much broader modeling program aimed at more complex and more realistic models, which also represented the other subsystems of the earth (Hasselmann 1979; for further development of this idea see Bretherton et al. 1986, 1988). Consequently, the Max Planck Institute initially focused on the development of a full ocean model before it entered atmospheric modeling in the mid-1980s (Hasselmann 2006).

Climate models as policy tools: Towards deterministic modeling

The SCEP and SMIC framed global climatic change as "one of the most important questions of our time," as Kellogg had written in 1971, and called for more definite predictive knowledge about future climate. Climate modelers of the first generation responded with limited engagement to that call. Manabe and Wetherald investigated climate change for doubled CO_2 in their paper from 1967; and they continued to serve interests in climate prediction, although for them it remained a subordinate interest. NCAR's Warren Washington only began to explore the issue from 1977 when he became involved in the new program of the Division of Carbon Dioxide and Climate Research of the US Department of Energy. Akio Arakawa did not pursue climate prediction, but made his model available to the Goddard Institute of Space Studies, which developed an interest in prediction.

It was a new generation of climate modelers who put a strong focus on predictive modeling, notably Stephen H. Schneider and James E. Hansen. They responded to William Kellogg's emphasis on the political urgency of the problem of climate change, pursued predictive ambitions, and compromised in model development and use with a strong element of pragmatism to produce politically relevant knowledge. Schneider was an engineering student at Columbia University when he started a first modeling experiment in 1970 with atmospheric scientist Ichtiaque Rasool of the Goddard Institute of Space Studies (where he also met James Hansen), using a simple one-dimensional radiative convective model. In his biographical recollections, Schneider explained that he knew little about these models and their deficiencies.

> Nonetheless, I was drawn to the power of the idea. We could actually simulate Earth's temperature and then pollute the model in order to figure out what might happen before we had polluted the actual planet.
>
> (Schneider and Flannery 2009: 21)

In April 1971, William Kellogg invited Schneider to serve as a rapporteur at the SMIC meeting in July. Shortly thereafter he invited Schneider to move to NCAR to join in their climate modeling efforts (Schneider and Flannery 2009: 23–27).

In a popular book published in 1976, Schneider expressed political concern about climate warming and emphasized the importance of climate models. He described the limitations and uncertainties of these models and the restrictions put upon simulation due to limited computer power, but argued for using these models anyway.

> Unfortunately, *for the task of estimating the potential impact of human activities on climate the models are just about the only tools we have.* Should we ignore the predictions of uncertain models? ... I think not—a political judgment, of course. ... Once we know reasonably well how an individual climatic process works and how it is affected by human activities (e.g. CO_2-radiation effect), we are obliged to use our present models to determine whether the changes induced by these human activities could be large enough to be important to society.
>
> (Schneider 1976: 147, 148)

Schneider described the dilemma climate scientists faced by referring to the metaphor of a fortune teller's crystal ball:

> The real problem is: If we choose to wait for more certainty before actions are initiated, then can our models be improved in time to prevent an irreversible drift toward a future calamity? [...] This dilemma rests, metaphorically, in our need to gaze into a very dirty crystal ball; but the tough judgment to be made here is precisely how long we should clean the glass before acting on what we believe we see inside.
>
> (Schneider 1976: 149)

Hansen was educated in physics and astronomy at the University of Iowa. In 1967, he joined the Goddard Institute for Space Studies (GISS) in New York as a postdoctoral researcher, where he specialized in theoretical work on radiative transfer in planetary atmospheres. At about this time, as Hansen later recalled, GISS director, Robert Jastrow, "concluded that the days of generous NASA support for planetary studies were numbered" and began to "direct institutional resources toward Earth applications" (Hansen et al. 2000: 128). Around 1970, Hansen joined a weather prediction modeling project based on Arakawa's general circulation model. This involvement, he continued, "made it practical for us to initiate a climate modeling effort several years later" (Hansen et al. 2000: 129). Hansen explained his interests and ambition in climate modeling in an interview with Spencer Weart, in which he distinguished between the development of climate models and their application. He described the difference between his style and that of climate modeling pioneer Akio Arakawa in the following way:

> He will always be in the design, I think. You know, if you want to do real applications, then you really have to just be willing to go ahead and do something [...] We're taking the model and using it for climate applications. It's hard to have enough time to work on the basic structure of the model and also use it.
>
> (Weart 2000)

In 1981, Hansen and coworkers published a landmark paper in *Science*, which was also based on simulation results of a one-dimensional climate model, simplifying the atmosphere to one vertical column (Hansen et al. 1981). In this paper the authors accurately listed problems and uncertainties. These included many factors such as: a lack of information on vegetation albedo feedback; a "lack of knowledge of ocean processes"; the "impact of tropospheric aerosols on climate is uncertain in sense and magnitude due to their range of composition"; and the "nature and causes of variability of cloud cover, optical thickness, and altitude distribution are not well known" (Hansen et al. 1981: 959–960). Despite many sources of uncertainty, the authors achieved a good agreement of calculated and observation-based data by assuming CO_2 concentration, aerosols

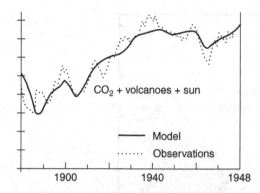

Figure 6.2 Observed and modeled global mean temperature changes.
Source: Hansen et al., 1981: 963.

from volcano eruptions, and variations of solar radiation as the main drivers of climatic change (see Figure 6.2). They considered this simple validation a sufficient means to create trust in model performance. Uncertainty was domesticated by providing an extensive qualitative discussion and ignoring it in the quantitative representation of simulation results. The authors concluded:

> The general agreement between modeled and observed temperature trends strongly suggests that CO_2 and volcanic aerosols are responsible for much of the global temperature variation in the past century. Key consequences are: (i) *empirical evidence* that much of the global climate variability on time scales of decades to centuries is deterministic and (ii) *improved confidence* in the ability of models to predict future CO_2 climate effects.
>
> (Hansen et al. 1981: 964; italics in original;
> for a more detailed discussion of Hansen's work
> see Heymann 2012)

After having established "improved confidence" in the model, Hansen and his co-authors proceeded to present future projections of global climate change based on various emission scenarios extending to the year 2100, about 120 years ahead (see Figure 6.3). According to these projections, global mean temperatures were expected to rise within a range of about 1°C in a scenario of slow or no economic growth with depleted oil and gas resources, or about 4.5°C if economic growth was fast. The authors were aware of the provisional character of such projections, because models "do not yet accurately simulate many parts of the climate system, especially the ocean, clouds, polar sea ice, and ice sheets." They also emphasized the possibility of changes in climate forcing due to factors such as a decrease in solar luminosity or volcanic eruptions which "may slow the rise in temperature." But they indicated confidence that "CO_2 warming should rise above the noise level of natural climate variability in this century" (Hansen et al. 1981: 964).

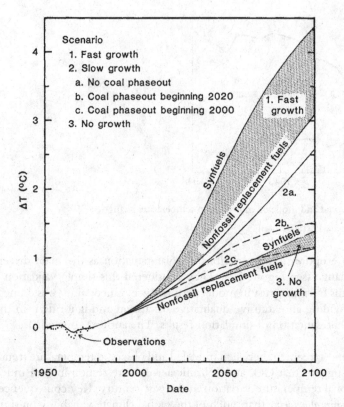

Figure 6.3 Climate projection to the year 2100.

Source: Hansen et al. 1981, p. 963.

Hansen's straightforward pragmatism and his far-reaching conclusions in spite of the tremendously simplified approach did not easily convince his peers. Hansen later told the story of this *Science* paper in an interview. He explained that he "had been working on it for a couple of years," but had problems getting the work published. "I sent it twice to Science and once to Nature and all these times, it got sent back" (Weart 2000). Once the revised paper was eventually accepted by *Science*, Hansen immediately planned for maximum publicity. He sent the accepted paper to renowned science writer Walter Sullivan, who published an article titled "Study finds warming trend that could raise sea levels" in the *New York Times* six days before Hansen's paper appeared in *Science* (Sullivan 1981). Hansen's paper and its publicity provoked a great deal of criticism. Atmospheric and climate scientists like Mike McCracken (1983) and Ichtiaque Rasool (1983) were quite upset and published responses; Rasool's amounted to "something like a personal attack" (Weart 2000). In his critique "On predicting calamities," Rasool wrote that the paper by Hansen et al. made him "reflect on the growing problem of how to separate good science from good fiction, especially when the press coverage picks up the latter" (Rasool 1983: 201). Hansen was also accused of trying to get funding by attracting public attention. But, in fact, he lost funds from the Department of Energy, which reversed a positive decision after critical responses to his paper were published (Hansen 2007: 2).

According to many scientists, Hansen moved too quickly, publishing results that were unduly confident and overly sensational. In one respect, however, his work reflected an increasingly shared—if not necessarily expressed—consensus among climate modelers: the climate system was assessed as deterministic enough to facilitate predictions of the second kind, and predictive modeling of long-term global climatic change was considered a top priority goal. This consensus received visibility and official political backing with the publication of the so-called "Charney Report" in 1979, which was commissioned by the National Academy of Science and prepared by a study group of nine leading climate scientists under the leadership of Jule Charney. Based on future projections of climatic change with Manabe's and Hansen's models, the study group cautiously concluded that it "finds no reason to doubt that climate changes will result and no reason to believe that these changes will be negligible" (Charney et al. 1979: xiii). While the report discussed weaknesses and problems of current models, it tacitly assumed the possibility of deterministic modeling without even mentioning the issue. Like Kellogg in his report, it assumed that warming might be delayed a little because of the interaction between the atmosphere and the ocean.

Conclusion: Seeking political relevance

Interest in long-term global warming and in investigating it with climate modeling was not a natural outcome of research trajectories in the study of climate. It was rather the result of complex and contingent historical processes and events. Interests in climate, questions related to climate and research approaches for the investigation of climate were diverse. Scientists had to compete for scientific and political attention. In an age of surging environmental concern, scientists like William Kellogg recognized future global warming due to increases of CO_2 levels in the atmosphere as an important scientific and political issue. These scientists identified the need and assumed the task of compiling the relevant knowledge on climate and its potential changes, formulating conclusions and warnings in accessible, strong language and raising public and political awareness. Their cause, however, was subject to competition for authority and resources.

Climate models could be used to generate policy-relevant predictions, a purpose Kellogg soon articulated as a new research priority. While modelers of the first generation generally had different ambitions and responded to his call only carefully or even reluctantly, a second generation of younger, more politically minded modelers such as Stephen Schneider and James Hansen took up the challenge. They developed not only the science of climate modeling from a heuristic into a predictive direction, which demanded new priorities and decisions, compromises with regard to the adequate representation of processes and a good deal of pragmatism; they also took great pains to promote and publicize their achievements and conclusions. Their work and their positions were controversial and not accepted and shared by all scientists. Still, they formed a foundation of an emerging culture of climate prediction, which only a few years later would largely define the domain of climate modeling, with the production of politically relevant predictive knowledge on climate change becoming its core task.

Predictive modeling was not easy to achieve and needed to gain scientific authority. It raised theoretical problems such as the problem of uncertainty or the even more fundamental question as to whether the climate system was transitive (deterministic), intransitive (chaotic) or almost intransitive (hopefully deterministic enough), a question which could not easily be settled, and which was rather ignored or set aside on the basis of plausibility arguments. By the late 1970s, Kellogg, Schneider, Hansen, and other American climate scientists took a pragmatic stance and regarded deterministic climate predictions of the second kind as feasible—despite objections by scientists like Lorenz and Hasselmann. This position was also introduced into politics through the US National Academy of Science and the WMO, institutions which furnished this approach with additional authority. At this point, plausibility arguments and a good deal of pragmatism and the experience that climate models seemed to work in useful and consistent ways seemed sufficient to domesticate inherent uncertainties. Uncertainties, however, persisted. They remained a paramount concern in climate prediction and soon demanded further efforts for their domestication (see also Landström in this volume).

The research trajectory shaped by these efforts represented quite a significant shift, because it prioritized the production of predictive knowledge over heuristic model use, dedicated to improving the understanding of atmospheric and climatic processes. In Hansen's words, it represented a shift in focus from "development" to "application" of climate models.

The impact was deep and long lasting. This shift bore the promise of resources, and of channeling these resources, for the efficient and quick delivery of politically useful data, while laborious basic research, which demanded much time and resources, received less scientific attention. Only recently, after disappointing stagnation of model performance in spite of decades of development and use, have climate scientists increasingly demanded that the focus be redirected back from model use toward improving understanding (Held 2005; Bony et al. 2011). The shift from heuristic to predictive modeling was a critical juncture, which from the 1970s on increasingly marginalized the heuristic approach and its ambition to expand and deepen the understanding of climate rather than producing long-term predictions.

References

Arakawa, A. (2000) "A personal perspective on the early years of general circulation modeling at UCLA," D. A. Randall (ed.) *General Circulation Model Development, Past, Present, Future*, San Diego, CA: Academic Press, pp. 1–65.

Bony, S, B. Stevens, I. H. Held, J. F. Mitchell, J.-L. Dufresne, K. A. Emanuel, P. Friedlingstein, S. Griffies and C. Senior (2011) "Carbon dioxide and climate: Perspectives on a scientific assessment," *WCRP Open Science Conference*, available from <http://www.wcrp-climate.org/conference2011/positionpapers.html>, accessed 30 April 2014.

Bretherton, F. P. et al. (1986) *Earth System Science: A Program for Global Change*, Washington, DC: NASA.

Bretherton, F. P. et al. (1988) *Earth System Science: A Closer View*, Washington, DC: NASA.

Broecker, W. S. (1975) "Climatic change: Are we on the brink of a pronounced global warming?" *Science* 189, pp. 460–463.

Charney, J. G. et al. (1979) *Carbon Dioxide and Climate: A Scientific Assessment*, Washington, DC: National Academy of Sciences.

Dahan-Delmedico, A. (2001) "History and epistemology of models: Meteorology as a case study (1946–1963)," *Archive for the History of the Exact Sciences* 55, pp. 395–422.

Dronin, N., M. Bellinger and G. Edward (2005) *Climate Dependence and Food Problems in Russia, 1900–1990*, Herndon, VA: Central European University Press.

Edwards, P. N. (1997) *Interview with Dr. Akio Arakawa, University of California, Los Angeles*, transcript available at <http://www.aip.org/history/ohilist/35131_1.html>, accessed 30 April 2014.

Edwards, P. N. (2010) *A Vast Machine: Computer Models, Climate Data, and the Politics of Global Warming*, Cambridge, MA: MIT Press.

Edwards, P. N. (2011) "History of climate modeling," *WIREs Climate Change* 2, pp. 128–139.

Frankignoul, C. and K. Hasselmann (1977) "Stochastic climate models, Part 2. Application to sea-surface temperatures and thermocline variability," *Tellus* 29, pp. 289–305.

Gates, W. L. (ed., 1979) *Study Conference on Climate Models, Performance, Intercomparison, and Sensitivity Studies*, (Washington, DC, 30 April 1978), Geneva, Switzerland: WMO.

Hansen, J. E. (2007) "Scientific reticence and sea level rise," *Environmental Research Letters* 2, pp. 1–6.

Hansen, J. E., D. Johnson, A. Lacis, S. Lebedeff, P. Lee, D. Rind and G. Russell (1981) "Climate impact of increasing atmospheric carbon dioxide," *Science* 213, pp. 957–66.

Hansen, J. E., R. Ruedy, A. Lacis, M. Sato, L. Nazarenko, N. Tausnev, I. Tegen and D. Koch (2000) "Climate modeling in the global warming debate," D. A. Randall (ed.) *General Circulation Model Development, Past, Present, Future*, San Diego, CA: Academic Press, pp. 127–164.

Harper, K. C. (2008) *Weather by the Numbers. The Genesis of Modern Meteorology*, Cambridge, MA: MIT Press.

Hasselmann, K. (1976) "Stochastic climate models, Part 1. Theory," *Tellus* 28, pp. 473–485.

Hasselmann, K. (1979) "Some comments on the design of model response experiments for multi-time-scale systems," L. W. Gates (ed.) *Study Conference on Climate Models, Performance, Intercomparison, and Sensitivity Studies*, (Washington, DC, 30 April 1978), Geneva, Switzerland: WMO, pp. 1037–1049.

Held, I. (2005) "The gap between simulation and understanding in climate modeling," *Bulletin of the American Meteorological Society* 86, pp. 1609–1614.

Heymann, M. (2012) "Constructing evidence and trust: How did climate scientists' confidence in their models and simulations emerge?" K. Hastrup and M. Skydstrup, (eds.) *The Social Life of Climate Change Models: Anticipating Nature*, New York, NY: Routledge, pp. 203–224.

Kellogg, W. W. (1971) "Predicting the climate," W. H. Matthews, W. W. Kellogg and D. Robinson (eds.) *Man's Impact on the Climate*, Cambridge, MA: MIT Press, pp. 123–132.

Kellogg, W. W. (1977) *Effects of Human Activities on Global Climate*, (Technical Note 156), Geneva, Switzerland: WMO.

Kellogg, W. W. and S. H. Schneider (1974) "Climate stabilization: For better or for worse?" *Science* 186, pp. 1163–1172.

Lemke, P. (1977) "Stochastic climate models, Part 3. Application to zonally averaged energy models," *Tellus* 29, pp. 385–392.

Lewis, J. M. (1998) "Clarifying the dynamics of the general circulation: Phillips's 1956 experiment," *Bulletin of the American Meteorological Society* 79, pp. 39–60.

Lewis, J. M. (2000) "Clarifying the dynamics of general circulation: Phillips's 1956 Experiment," D. A. Randall (ed.) *General Circulation Model Development, Past, Present, Future*, San Diego, CA: Academic Press, pp. 91–125.

Lorenz, E. (1968) "Climatic determinism," *Meteorological Monographs* 8(30), pp. 1–3.

Lorenz, E. (1971) "Climatic change as a mathematical problem," W. H. Matthews, W. W. Kellogg and W. D. Robinson (eds.) *Man's Impact on the Climate*, Cambridge, MA: MIT Press, pp. 179–189 (originally published in: *Journal of Applied Meteorology* 9(3), 1970, pp. 325–329).

MacCracken, M. C. (1983) "Climatic effects of atmospheric carbon dioxide," *Science* 220, pp. 873–874.

Manabe, S. and R. T. Wetherald (1967) "Thermal equilibrium of the atmosphere with a given distribution of relative humidity," *Journal of the Atmospheric Sciences* 24, pp. 241–59.

Matthews, W. H., W. W. Kellogg, and G. D. Robinson (eds., 1971) *Man's Impact on the Climate*, (Collected background papers of SCEP), Cambridge, MA: MIT Press.

Mintz, Y. (1965) "Very long-term global integration of the primitive equations of atmospheric motion," WMO (ed.) *WMO-IUGG Symposium on Research and Development Aspects of Long-Range Forecasting, Boulder, Colorado, 1964, (Technical Note No. 66),* Geneva, Switzerland: WMO, pp. 141–155.

Mitchell, J.M. Jr. (1977) "Carbon dioxide and future climate," *Environmental Data Service* 3, pp. 3–9.

National Academy of Sciences, Committee on Atmospheric Sciences Panel on Weather and Climate Modification (1966) *Weather and Climate Modification: Problems and Prospects, (2 vols.),* Washington, DC: National Academy of Sciences.

Nebeker, F. (1995) *Calculating the Weather. Meteorology in the 20th Century*, San Diego, CA: Academic Press.

PSAC The President's Science Advisory Committee (1965) *Restoring the Quality of Our Environment. Report of the Environmental Pollution Panel*, Washington, DC: The White House.

Rasool, S. I. (1983) "On predicting calamities," *Climatic Change* 5, pp. 201–202.

SCEP Study of Critical Environmental Problems (1970) *Man's Impact on the Global Environment: Assessment and Recommendation for Action*, Cambridge, MA: MIT Press.

Schneider, S. H. (1976) *The Genesis Strategy: Climate and Global Survival*, New York, NY: Plenum Press.

Schneider, S. H. and T. Flannery (2009) *Science as a Contact Sport, Inside the Battle to Save the Earth's Climate*, Washington, DC: National Geographic.

Shackley, S., P. Young, S. Parkinson and B. Wayne (1998) "Uncertainty, complexity and concepts of good science in climate change modeling: Are GCMs the best tools?" *Climatic Change* 38, pp. 159–205.

SMIC (1971) "Study of man's impact on the climate," *Man's Impact on the Climate*, Cambridge, MA: MIT Press.

Sullivan, W. (1981) "Study finds warming trend that could raise sea levels," *New York Times*, 22 August 1981.

UNCHE (1973) *Report of the United Nations Conference on the Human Environment*, Stockholm, Sweden: UNCHE.

Washington, W. M., A. J. Semtner, G. A. Meehl, D. J. Knight and T. A. Mayer (1980) "A General Circulation Experiment with a Coupled Atmosphere, Ocean and Sea Ice Model," *Journal of Physical Oceanography* 10, pp. 1887–1908.

Weart, S. (2000) *Interview with James Hansen*, (Center for History of Physics. College Park: American Institute of Physics), transcript available from <http://www.aip.org/history/ohilist/24309_1.html>, accessed 30 April 2015.

Weart, S. (2010) "The development of general circulation models of climate," *Studies in History and Philosophy of Modern Physics* 41, pp. 208–217.

Weart, S. (2014) "General Circulation Models of Climate," *The Discovery of Global Warming*, (A hypertext history of how scientists came to [partly] understand what people are doing to cause climate change), available from <http://history.aip.org/climate/GCM.htm>, accessed 14 February 2017.

White, R. M. (1979) "Keynote address," *Proceedings of the World Climate Conference, a conference of experts on climate and mankind*, 12–23 February 1979, Geneva, Switzerland: WMO.

WMO (1975) *The Physical Basis of Climate and Climate Modeling*, (Report on the Study Conference on the Physical Basis of Climate and Climate Modeling, Stockholm 1974), Geneva, Switzerland: WMO.

7 How to develop climate models? The "gamble" of improving climate model parameterizations

Hélène Guillemot

Introduction

How should predictions of the future climate be improved, and confidence in those predictions increased? What is the way forward for computer models, key instruments of the climate sciences and the only accepted tools for predicting climate changes? The principal driving force behind climate modeling seems to be the exponential growth of computing power, allowing models to include ever greater numbers of physical, chemical, and biological components, with growing spatial resolution, to provide projections of climate change at a regional scale (see also Mahony in this volume). However, this trend is not hegemonic, and in the climate sciences community, the debate about strategies for model development rages on.

Indeed, despite major progress in computing and observation, the precision and reliability of climate change predictions have changed little over the last decade. Climatologists have long known that the flaws and uncertainties of simulations are mainly attributable to their representations of sub-grid scale physical processes such as clouds and convection, known as parameterizations. These parameterizations thus lie at the heart of animated debate: are they in a "deadlock" (Randall et al. 2003)? Why have developments in modeling been failing to lead to corresponding improvements in prediction?

In this chapter, I will rely on the development of a new parameterization—the representation of convection and clouds—of the climate model of the Laboratoire de Météorologie Dynamique (LMD) of the Centre National de la Recherche Scientifique (CNRS), in Paris, to shed light on the practices, choices and points of view of LMD researchers and other modeling groups with regard to model development. The first section exposes the discussions on parameterizations versus other strategies for model development; the second section looks at the design, construction, and validation of the new parameterization of the climate model of the LMD; finally, in the third section we will see that these different approaches involve varied epistemic conceptions of models, and of their roles, uses, and limits, grounded in different practices and institutions. In all, this chapter demonstrates how the existence of different "cultures of prediction" fundamentally shapes the courses of climate model development.

Debates on the evolution of climate modeling

For some years, there has been animated debate on the development of models: the World Climate Research Program (WCRP) has organized numerous workshops and conferences on this topic; reports and articles have been published by researchers and groups seeking to advance their own visions of the direction that climate modeling should take (Jakob 2010; Jakob et al. 2010; Shukla et al. 2009; Knutti 2010; Randall et al. 2003; Bony et al. 2011; Held 2005). To grasp what is at stake, a look at the structure of climate models and the role of parameterizations is needed. ·

The dual structure of the Global Circulation Models

Atmospheric models—be they climate models or numerical weather prediction models—simulate planetary-scale atmospheric flow as represented in a three-dimensional grid; this is why they have been called Global Circulation Models (GCM). For each grid cell and at each time step, the computer calculates the value of the variables characterizing the mean state of the atmosphere within the cell (temperature, wind, humidity, etc.) from their values at the previous time step, by calculating the algorithms that constitute the model. Atmospheric models are organized into two parts: A "dynamic core," which describes the movements of air masses at the grid scale using algorithms which discretize the laws of fluid dynamics (Navier–Stokes equations); and a so-called "physics" part, which deals with the physical processes that influence weather or climate occurring at various scales below the grid scale (the grid size in a climate model is about a hundred kilometers). These sub-grid scale processes are represented by "parameterizations" describing their statistical impact on large-scale variables.[1]

The dynamic core of the model represents the system of the global atmosphere, which obeys fundamental laws (conservation of energy, mass, water, etc.) on a large scale. With parameterizations, however, additional theories and formalisms are incorporated into climate models (Galison 1996). Thus, climate (or weather) models rest on a decoupling of explicitly calculated large-scale dynamic and small-scale phenomena whose effects must be represented statistically. GCMs are deeply structured by this duality, which has subsisted since their inception, although they have changed considerably in size and complexity over time.

Parameterizations are extremely diverse: some are veritable physical models within the model, with their own internal variables and their own equations, others are more empirical. From the very beginning of weather and climate modeling, atmospheric physicists have identified essential subscale processes responsible for the exchange of energy, momentum, and/or water between the atmosphere and the earth surface: radiative transfer, small-scale turbulence transporting energy in the boundary layer (the layer of air in the immediate vicinity of the surface), convection on a larger scale, clouds, precipitation, and so on. These processes were considered in the first models through simple parameterizations, which have since been greatly refined. Broadly speaking, these improvements consist in replacing fixed or "tuned" parameters with sub-models of physical processes (which themselves include empirical or tuned parameters, at a lower level). For example, the first climate models represented convection as a large-scale adjustment, transporting energy and humidity toward the upper atmospheric layers and eliminating water beyond a saturation threshold, through condensation and precipitation. In later models, convection has been described on more physical grounds, and its interactions with other processes represented.

Over the course of their half-century of history, climate models have been developed both through improvements in these main parameterizations, and through the integration of new components to include more milieus and phenomena: the ocean, sea ice, vegetation, ecosystems, atmospheric chemistry, carbon cycle, marine biogeochemistry, and so on—thus becoming "Earth System Models." During the 1990s and 2000s, core physical parameterizations were somewhat neglected in favor of the complexification of models and coupling with other components of the earth system (Dahan 2010). Some climate researchers have criticized this "overemphasis on new components that increase complexity [...] often without addressing existing big or longstanding problems" (Jakob et al. 2010: 3, 4). But in the past several years there has been a growing awareness of the need to improve the core of the models.

Persistent biases and need for better core parameterizations

Despite substantial progress in many domains, the range of uncertainty on climate sensitivity[2] has not decreased since the first Intergovernmental Panel on Climate Change (IPCC) reports (1.5C° to 4.5C°), and there is relative stagnation with regard to the main flaws in climate models: to take just one example, little is known about how to simulate precipitation in tropical regions and how it would change in a warmer climate. Climatologists know that many of these recurring flaws are due to the poor representation of sub-grid-scale processes, of clouds and convection in particular. As these processes are represented differently depending on the model, the parameterizations of clouds and convection constitute the main sources of uncertainty for projections of the future climate.

Yet, developments in climate science seem to provide opportunities for the improvement of core physical parameterizations. Nowadays they have been recognized as critical for the quality of simulations and hence for research on the impacts of climate change: all components of the climate depend on the representations of main processes (no good representation of vegetation dynamics or carbon cycle without a good distribution of precipitation). The improvement of parameterization is also favored by the current availability of an unprecedented range of observations—ground-based, space-borne and in situ campaigns—to the point that some consider that we live in a "golden age" of observations.

Moreover, for 20 years, international research programs have been dedicated to improving GCM parameterizations using knowledge acquired through detailed modeling of clouds and convection. High resolution numerical modeling saw considerable growth in the 1980s; these limited area models explicitly calculate the vertical convection movements within the atmosphere, which in GCMs, in contrast, are parameterized.[3] Two types of small scale models are distinguished: Cloud Resolving Models (CRM; grid cells on the order of a kilometer, domain of 100 to 1000 km) and Large Eddy Simulations (LES: grid cells 10 to 100 m in size, domain of 10 to 200 km). LES and CRM models have made it possible to study and simulate all categories of clouds in many types of climate, as well as their collective organization (squall lines, storms, etc.). The development of high-resolution models benefited from large-scale, open-air experiments organized over the last 40 years, involving several international teams and instruments transported by boat, airplane, balloon and satellite, in order to observe a particular atmospheric process (storms in the Atlantic, the West African monsoon, etc.). Beginning in the 1990s, international programs like EUROCS (European Project On Cloud Systems) and GCSS (GEWEX Cloud System Study)[4] have used these field campaigns to help improve cloud parameterizations, seeking to foster collaboration between the global model community and the process community: "Observations from field programs will be used to develop and validate the cloud-resolving models, which in turn will be used as test-beds to develop the parameterizations for the large-scale models" (GCSS Team 1993: 387).

However, although these observations and research programs have indeed improved the understanding of processes, they have more rarely led to major revisions in the parameterizations of models. Scientists involved in that domain advance several reasons for this limited improvement. The considerable effort and time required to conceive a new parameterization (as we will see in next section) does not encourage reopening long-established parameterizations to revision. Some researchers invoke the small number of "developers" as compared to the number of "users," and also the lack of incentives: despite established discourse and programs, the issue of climate change draws research toward regional predictions and impacts more than toward the development of parameterizations[5]. Moreover, this type of activity is reputed to be academically unrewarding: it leads to fewer publications than numerical experiments or simulations, is less valued than opening up new domains—and fits less well with current research funding criteria. As an LMD researcher (L1)[6] explained:

> It's difficult to justify working on old questions. If you don't get the results you need, you have to change topics [...] When a problem is new, the system is poorly constrained, you get strong answers; so that gives you papers that are easier to write [...].
>
> (L1, pers. comm., July 2012)

These model developers thus tend to consider themselves a minority, facing the dominant trends in modeling which promote other scientific practices.

The gap between processes and climate

Climate model developers express a disconnection between process studies on the small scale and climate model assessment on the large scale. This young researcher (L2), for instance, senses a lack of communication and of activity between these communities:

> I went to a GCSS conference in Boulder on process studies, fascinating. I asked: where do we talk about the impact that parameterizations have in models, at a large scale and in the long term, on the climate, and about why the models aren't improving? The answer was: it will be talked about in the workshop on systematic model errors in Exeter. I went there. They were looking at what the climate biases were, but this still wasn't the place where we would talk about the question of where these biases come from [...] But it's not an easy question, we don't know how to do that. You get the impression that there's a lack of activity between these two communities.
>
> (L2, pers. comm., December 2013)

Some researchers insist on the need to strengthen links and coordination between the process and climate areas. But the problem is deeper and touches on a fundamental characteristic of climate modeling: the difficulty of attributing the characteristics of simulations to components of the model, because of the multitude of processes and interactions at many scales. This problem is nicely summarized in a WCRP white paper: "Model development is hindered by a lack of understanding of how a poor representation of cloud scale processes and cloud scale dynamics contribute to model biases in the large scale circulation features and influence future projections" (WCRP 2010).[7] This "gap between processes and large-scale climate" (in the climatologist's parlance) has been characterized by philosophers as a consequence of the "epistemic opacity" of computer simulation (Humphreys 2004: 147) or their "confirmation holism" (Lenhard and Winsberg 2010: 254).

As an instance of this epistemic opacity, even when a new parameterization successfully reproduces an aspect of the current climate, there is no guarantee that the same will be true for the future climate, because the influence of the numerous physical processes may vary in unknown ways over different time-scales. That is why, according to certain modelers, improving parameterizations is a "gamble".

Superparameterization and super-models

Given the difficulty of this problem, certain climatologists prefer not to take the gamble and advocate other paths for the development of models. In an article entitled "Breaking the cloud parameterization deadlock," four renowned specialists in cloud physics, considering the considerable amount of complex knowledge on cloud processes and their poor representation in climate models, claim that it is impossible "to parameterize all of this complexity with quantitative accuracy" (Randall et al. 2003: 1551). As "our rate of progress is unacceptably slow" they propose a "new and different strategy" called "super-parameterization," a sort of hybrid between a parameterization and an explicit calculation that consists in replacing the cloud parameterization in each grid cell with a high-resolution two-dimensional CRM-type model (Gramelsberger 2010). Another strategy was advanced at the World Modelling Summit for Climate Prediction in 2008, and taken up in a "declaration" entitled *Revolution in Climate Prediction is Both Necessary and Possible* (Shukla et al. 2009). This text recommends the launch of a "World Climate Prediction Project," equipped with computers at least 1000 times more powerful than those currently available, to develop models at kilometer resolution, capable of explicitly calculating convection and even boundary layer eddies—in other words, extending LES-type detailed models to the entire planet. These super-models, including numerous biogeochemical processes, could provide "operational climate prediction at all time scales, especially at decadal to multidecadal lead times"—what is called "seamless prediction," merging weather forecasting and climate prediction.

This colossal project provoked rather heated debate in the climate modeling community. Several types of objections have been raised: These models will not be available for several decades; even high-resolution models contain parameterizations, at smaller scales; and if the idea of seamless prediction is seen as interesting, since there is no scientifically relevant boundary between modeling weather and climate, some climate scientists "remain skeptical about the overall benefit of the 'one size fits all' model" (Knutti 2010: 401), because the objectives and ways of working with weather or climate models are different. Another sensitive topic is the risk of a hegemony of international super-models, favoring certain objectives and strategies. The stakes in this "End of model democracy" (Knutti 2010: 395) are both political and scientific, involving the diversity of approaches, the autonomy of scientific decision-making and the quality of forecasts.[8] Finally, very high-resolution models are not guaranteed to produce good climate simulations: there are discussions on the appropriate temporal and spatial scales for these models.

The overhaul of a parameterization at the heart of the model

The researchers at the LMD are among those who took the gamble on parameterization. For more than a decade, a few scientists have developed a new parameterization that was implemented in the LMD atmospheric model in 2011.[9] This new version of the model was ready in time to participate (alongside the old version) in the fifth phase of the Coupled Model Intercomparison Project (CMIP5), and thus to be included in the latest report of the IPCC.[10]

Representing surface turbulence, tropical storms,
and fair-weather cumulus

Rather than a single new parameterization, it is a new set of parameterizations that has been developed and recently implemented in the LMD model. The representations of several atmospheric phenomena are now combined: turbulence in the boundary layer (up to an altitude of 10 to 100 m), high-altitude convection with the associated enormous cumulonimbus clouds (so-called "deep convection"—10 to 20 km), and also an intermediate-scale phenomenon that was not described in previous models: convective movements in the boundary layer (1 to 3 km in altitude), called "thermal plumes," which give rise to the small fair-weather cumulus clouds.

The representation of deep convection, in particular, has been modified profoundly. The decision to change this parameterization was motivated by a recurring flaw in the LMD model (and in most GCMs): the precipitation time lag in the tropics. In the models, the rains began at noon, whereas in the real world they tend to fall in the late afternoon—a serious bias, which revealed a poor representation of the phenomenon and which had serious consequences for the energy balance in the tropics. The LMD researcher who worked on tropical convection (L3) came to an understanding of the origin of this flaw in 2000, in a discussion with a researcher from Météo-France (the French center for weather forecasting) who was a specialist in the study of clouds at a small scale. The time lag is due to cloud-internal phenomena, which had already been observed and studied by climatologists, but not yet taken into account in global models. It took more than ten years for L3 to achieve the representation of these phenomena in a parameterization (—which includes processes named "cold pools" and "density currents," which we will not describe here (Hourdin et al. 2013: 2193).

The new parameterization is the product of a partly individual and partly collaborative work. The first task is conceptualization: parameterizations do not represent all processes, but only those that scientists consider to play an important role (in our case, thermal plumes and cold pools, for example). Here the work consists in designing an idealized representation of the phenomenology of the processes and formulating these mechanisms as equations. Interactions between different processes must be represented as well, and the parameterization must be linked to the large-scale dynamics. The new parameterization of the LMD model represents "a radical change," according to of one of its creators (L4), because it is based on detailed studies of the actual physical processes: "this coupling between cold pools [...] and convection [...] allows for the first time to get an autonomous life cycle of convection, not directly driven by the large scale conditions" (Hourdin et al. 2013: 2198).

Collaborations between large- and small-scale modeling

Developing the new physics involved several researchers at the LMD, and gave rise to several PhD theses, but also relied extensively on a collaboration with the scientists at Météo-France's research center who study cloud processes at the small and medium scale by developing and using high-resolution atmospheric models. This collaboration is part of the worldwide dynamic mentioned above, aiming to help in developing parameterization in large scale models by using field campaigns and high-resolution models within the framework of international programs like EUROCS or GCSS.

However, moving from local case studies to a parameterization that is valid on the global scale raises fundamental difficulties. L3 recounted that he had been greatly impressed by the presentation of a high-resolution simulation at a conference in 2008. The speaker showed an animated simulation of convection and tropical storms over 24 hours within a 200 km-wide "box" at a resolution of 100 m.

> The room was silent, everyone was in awe. You could see lots of little density currents meeting each other, interlocking over the continents, in squall lines, fusing, colliding and teeming over the ocean [...]. The conclusion of the speaker as to the possibility of representing parametrically this high resolution simulation was: 'It's not manageable!'.
>
> (L3, pers. comm., August 2012)

Yet L3 tried to take up the challenge of moving from a small-scale, single-day simulation to a large-scale vision. As he explained:

> A parameterization has to be valid for the whole world: we start from the principle that all the storms in the world have density currents. You have to fit an average vision onto the explicit simulations, describe populations of cold pools and density currents statistically.
>
> (L3, pers. comm., August 2012)

High-resolution models are used in conceptualizing a parameterization, but also in validating it (Guillemot, 2010), following a methodology that is now well established. The modelers select case studies representative of different aspects of the phenomenon to be parameterized (convection above the ocean, squall lines etc.). For each case study, they perform a simulation with a Cloud Resolving Model at a resolution of 10 km. This high-resolution simulation is initialized and validated by field measurements, and then used as a reference to test the parameterization. To do this, the parameterization is incorporated into a one-dimensional version of the GCM (known as a "single-column model"), which consists of a single horizontal grid cell with all of the vertical layers above it. The exchanges with the exterior provided as "forcing" (heat, humidity) are the same for the single-column model and for the high-resolution model. The simulation drawn from the parameterization can thus be compared to the high-resolution simulation.

As a single-column model is much less computationally intensive than a GCM, while containing its entire physics, it makes it possible to test the parameterization on several case studies, since "to be qualified for 3D climate modeling, a parameterization must be valid both over ocean and continents, from the poles to the equator, on deserts or wetlands" (Hourdin et al. 2013: 2200). Then the parameterization is evaluated using the whole model.

A new parameterization creates disruptions

The final step is the implementation of the new parameterization in the model; it is also one of the most difficult. Indeed, parameterizations developed through process studies have often initially produced poor climate simulations. This apparent paradox is explained by "compensating errors": even if the previous parameterization was inferior to the new one, the model may have been tuned so that its flaws were compensated for by errors in other parameterizations of the model, producing the semblance of a correct climate. Modifying a parameter alters this balance by bringing new feedback effects into play. Thus, as a Météo-France climatologist explained, parameterizations "develop collective behaviours," which contribute to making the model "the expertise of a group." It is therefore essential that the different parameterizations of a model be coherent, with sectioning and process representations that are well harmonized. This is why, after the design work, the development of a new parameterization requires a long phase of adjustment and tuning.

What change in climate simulations did the new parameterization of the LMD model ultimately produce? What are the differences with respect to the old parameterization? The changes are substantial, which is not surprising since the processes represented affect all aspects of the climate and have an influence on large-scale circulation. Improvements were observed, but there was also deterioration in some points. The new parameterization clearly improved the aspects on which efforts had been focused, in particular the representation of boundary layer clouds and the variability of precipitation; and there is no longer a lag in the cycle of tropical precipitation. However, the new version of the model presents the flaws which are common to most coupled ocean-atmosphere models, and in some cases even amplifies these biases (shifted rainy areas, for example)

An interesting result is that the new and old parameterizations lead to markedly different sensitivities to increasing CO_2. While the standard version of the LMD model has a high sensitivity (relative to the mean of coupled GCMs), with the new physics the sensitivity is among the lowest. This difference needs to be interpreted in detail, but it doesn't come as a surprise, since the new parameterization substantially modifies the representation of low clouds, which are known to play an important role in climate sensitivity.

Conceptions of climate models and laboratory cultures

We have seen that modeling groups have different discourses, practices and strategies about the development of climate models. Now I will attempt to better characterize these conceptions, and to clarify what the oppositions really reveal.

Physical understanding, partitioning, hierarchizing

The development of the new parameterization of the LMD model is rooted in a "thought style" (Fleck 2008) shared by scientists, which is also acquired by young researchers who are socialized within the group, as one of them (L2) recounted:

> At the LMD there's quite a strong lab culture. Even when they're doing different things, they have a common way of seeing things, an approach that's very much based on the understanding of processes, more than elsewhere. [...] I was immersed in it, I became imbued with it. And I also made my own contribution.
>
> (L2, pers. comm., December 2013)

As L2 aptly noted, an essential component of the LMD common culture is the importance attached to physical understanding. This conviction was expressed in a position paper presented at a WCRP's conference in 2011 (Bony et al. 2011) written by nine climatologists including researchers from the LMD. The position paper analyses the "Charney report" (Charney et al. 1979 – a famous report on the effects of increasing CO_2 written by a group headed by Charney) and claim that the reason why this assessment was so "amazingly prescient" 35 years ago is "the power of its scientific approaches," namely "the emphasis on the importance of physical understanding gained through the use of theory and simple models" (Bony et al. 2011: 393).

Understanding is sometimes seen as difficult, impossible, or irrelevant in computer modeling: for the philosophers cited above, the "epistemic opacity can result in a loss of understanding" (Humphreys 2005: 147) and "the holism makes analytic understanding of complex models of climate either extremely difficult or even impossible" (Lenhard and Winsberg 2010: 253). However, for these climatologists, physical understanding is even more necessary in climate modeling. Moreover, it is the precondition of progress in the prediction of climate change: since the future climate cannot be observed, "confidence in our predictions will remain disproportionately dependent on the development of understanding" (Bony et al. 2011: 404). Thus these scientists assert a convergence—rather than an opposition—between the objectives of knowledge and prediction (see also Heymann and Hundebol in this volume), and they construct research questions so as to align both concerns.

This common culture manifests itself in discourse and practices, but also in material devices and the organization of the laboratory. Since the late 2000s, the LMD has maintained a panoply of simplified, up-to-date versions of the global model—atmosphere-only models, "aqua-planet" models (entirely covered in water), two-dimensional models, single-column models, and so on—which have simpler climatic systems and fewer variabilities than the GCM (no seasons, no monsoons, no El Niño, no dynamics, etc.), while preserving essential elements. The researchers use this spectrum of models of different complexities at the two "ends" of modeling: in the development of parameterizations (as we saw in the first part) and for overall analyses of simulations—for example, in order to identify "robust" climatic mechanisms, which are found in all configurations, even very simple ones (Stevens and Bony 2013).

Two approaches that play a major role in the practices and discourses of LMD researchers are the hierarchizing of priority and the partitioning of problems and phenomena. They involve determining what the main climatic mechanisms in simulation analyses are, what elements play an essential role and must be studied in priority; for example, what type of clouds are mainly responsible for model uncertainties (Bony and Dufresne 2005); and, as we saw in the previous section, which processes it is important to represent in parameterization. The LMD's new parameterization provides several cases of partitioning: The representation of "thermal plumes" in the boundary layer, for instance, led modelers to distinguish three different scales and to reconsider the boundaries between processes—boundaries that do not exist in detailed modeling, where the transition from cumulonimbus to cumulus clouds is continuous.

These modeling practices—establishing a hierarchy of priorities, identifying the principal factors and partitioning problems into more tractable subproblems—can be seen as ways of circumventing the entanglement of climate processes. Thus for these researchers, the characteristic holism of climate models, rather than making understanding impossible, reinforces the need to invent sophisticated methodologies to work around it (Guillemot 2014).

Parameterization as Achilles' heel or heuristic tool

How can the different approaches to model development outlined above be characterized? In the article on the "Cloud Parameterization Deadlock" (Randall et al. 2003), the authors find it

> ironic that we cannot represent the effects of the small-scale processes by making direct use of the well-known equations that govern them [and that] cloud parameterization developers [...] are trying to compute the statistics of these processes that matter for the large-scale circulation and climate, without directly representing the cloud processes at their 'native' space and time scales.
>
> (Randall et al 2003: 1548)

For these authors, there is something scandalous about representing cloud processes using parameterizations instead of using "the basic physical equations in which we have the most confidence [...] at their 'native' space and time scales." Likewise, it strikes them as almost shocking to separately represent phenomena which are not separated in reality.

The vision that emerges here is of a climate the entangled processes of which are governed by the same fundamental laws, and which must be represented in accordance with its real nature, in a continuous and unified fashion.

While the dynamic core based on physical laws is the most solid part of climate models, parameterizations, which are incomplete and error-prone, are seen as their "Achilles' heel"—a mere lesser evil to be used until computers sufficiently powerful to solve the equations of fluid dynamics at a small scale become available.

But this is not the only way of seeing the issue. First, calculating atmospheric phenomena from the laws that govern them is not necessarily the ultimate goal of this research. To certain climatologists, on the contrary, it can represent the renunciation of a deeper understanding of processes, as a scientist from Météo-France explained (with regard to the modeling of ocean circulation):

> If you take a model with 2-kilometer resolution, you get beautiful results, which match the observations [...]. But it's frustrating: if we have to solve everything explicitly by computer, that means we haven't understood the physics behind it [...]. We know that sub-grid activity is important in controlling general circulation, but we don't know how to write a trans-formation law.
>
> (scientist from Météo-France, pers. comm, July 2001)

Despite its "beautiful results," this researcher finds something "frustrating" about explicit resolution compared to "transformation laws," which express greater understanding. Likewise, discourse and practices of the LMD research-ers highlight the heuristic role of parameterizations. Rather than emphasizing continuities and unified representations, they emphasize the partitioning of problems, the identification of essential phenomena, the representation of the main processes. "Parameterizing means building the equations and modeling to see if we've understood; if we've taken the key ingredients into account," according to (L2). A parameterization "breaks down the problem and offers an interpretative framework" (L4, pers. comm, July 2012). With detailed explicit models, in contrast, it is more difficult to grasp what processes are at work: "You can't 'unplug' a mechanism to test hypotheses," unlike with a param-eterization, which distinguishes the different processes and coupling variables (L4, pers. comm., July 2012). In brief, parameterizations "sum up our under-standing of the system" (L2, pers. comm., December 2013).

That is why some LMD modelers were disappointed by a sort of "renuncia-tion" expressed by some of the most renowned scientists working on cloud parameterization. L2 recounted the dismay she experienced at a workshop on climate models where she had hoped to discuss physical questions with emi-nent colleagues:

> The most shocking thing was the prevailing discourse: 'Developing parameterizations is too risky, it's difficult; it shouldn't be done.' But that's exactly what should motivate us, that's why I did it: it's open, it's a challenge!.
>
> (L2, pers. comm, December 2013)

Confidence in the capacity to pick out the essential

If we try to further specify dividing lines in this area on the basis of the differences that have been highlighted, it emerges that what separates scientists is their confidence in the possibility of picking out the essential processes. Some—including the LMD researchers quoted above—argue that it is possible to identify these elements, using idealized models as an interpretative framework.

This conviction is anchored in a certain vision of the climate: although the climate is complex, it presents patterns, organized behaviors. Despite non-linearity, variability, and the number of factors and processes at work, "it's not just noise, there's an organization to it. There's an object to be studied, we're not aimlessly adding processes," an LMD researcher said (L6, pers. comm, July 2001). The atmosphere exhibits "dominant modes," and the job of the climatologist is to identify, explain and reproduce these "simple" structures or behaviors, which requires making use of both simple and complex models. As L6 explained:

> In a certain way nature chooses dominant modes, and we need complex models to go understand how and why [...] But the understanding we get with complex models isn't really well established until we've been able to take it as far as building a conceptual schema. Starting from that complexity, you have to manage little by little to draw out a few simple ideas about how the thing that you're studying works, something you can explain in words [...] And if you master the chain of links between simple and complicated, you think you understand things.
>
> (L6, pers. comm., July 2001)

In other words, given that "nature does a simple thing in a complicated way" (Arakawa 2000: 53) these modelers think it is possible to focus on relatively simple aspects of nature's behavior to try and understand that apparent simplicity and reproduce it by modeling it.

Other scientists, on the contrary, are convinced that the complexity of processes and their interactions condemns these attempts at building conceptual schemas. Their view is that "GCMs try to mimic nature's own complicated way of doing simple things" (Arakawa 2000: 53), the model is seen more as a black box, and the work of simplification is delegated to the computer. The search for understanding is not located at the level of parameterizations and their effects, but downstream, at the level of regional climates, impacts, and so on.

To put it in terms of confidence (and with some exaggeration): Some have confidence in the capacity of climatologists to represent the essential, while others do not. And the latter have confidence in the capacity of complex computer models to approach real climate and predict climate change—which the former doubt, as this LMD researcher (L5) expressed:

> Since the beginning of modeling, there's been this idea that we're going to develop GCMs, increase the resolution, etc., and that naturally we'll

approach the solution, the truth. And people realize that it's a long road, really long. [...] But it's an approach that's still deeply rooted for us—an engineering approach perhaps. [...] I think that that's not how we're going to get there [...]. The way we're going to be able to anticipate something like the solution is much more by understanding what's happening.

(L5, pers. comm., July 2012)

Still, the possibility of identifying dominant mechanisms within processes, or "simple" relations between climate and processes, raises questions. Can such mechanisms always be isolated? Can the necessary approximations actually be made? The answers will come from the practices of modelers. It may be hypothesized—although that would require further argument—that the capacity to extract dominant processes or to partition phenomena in a useful way depends less on the intrinsic properties of the climate than on cognitive expectations and the way in which questions are asked (Ruphy 2003).

Conclusion: Multi-scale practices and cultures of modeling

In this chapter, I did not attempt to describe the whole range of positions on the advancement of climate models—that would require a broader study of different modeling groups. Rather, on the basis of the work of modelers in a French laboratory who were developing a new physical parameterization, I tried to draw out some of the factors involved in shaping modeling choices. We have seen here that epistemic conceptions of climate modeling are closely tied to scientific practices: visions of the climate, its complexity, and its predictability are inseparable from the methods and instruments used to study it. These visions, practices, and "epistemic lifestyles" of modelers depend on numerous social and institutional factors, including disciplinary background, research goals, available tools, funding, academic collaborations and research cultures (Shackley 2001). Strategies for model development cannot be understood independently of these factors.

Among the various social factors involved in scientific choices, the institutional status plays a decisive role. In the case of the LMD, the status of the CNRS was important in making the completion of the new parameterization possible, by allowing scientists to work for several years on a parameterization that they considered important, in a time when this orientation was not considered a high priority—even if it meant, when necessary, to sacrifice publications for some time. These researchers could do without project-based funding, which favors other research pathways. L3, who worked full time for more than ten years on this parameterization, called this "an extraordinary luxury." As L4 declared: "Creating a new parameterization of the boundary layer is unfundable—you have to be at the CNRS to have time to do that kind of thing!"

These choices evidently would not have been possible in all institutional contexts. Researchers propose strategies and adopt "thought styles" (Fleck 2008) that are in line with expectations within their institution. Rivalries between types of institutions, which have their own instruments, modes of operation, ways of working, and visions of the sciences, play an important role in this context. In climate modeling, it is well known that there is competition between research laboratories and operational prediction centers: The choice of decadal prediction or the "seamless" strategy is often made by researchers working in weather forecasting centers who want to make use of their know-how and expertise from this domain. These scientists anticipate the expectations of policy makers, who are willing to base their decisions on high-resolution simulations at a decadal or multi-decadal scale (with regard to adaptation, for example; see also Mahony in this volume). This co-construction is encouraged by funding sources that are inclined to favor the most "technophilic" solutions.

As we saw, the work of modeling involves collaboration on several scales: Collaborations with a few researchers from related fields who contribute their expertise; collective work on, around, and with the model, requiring rigorous management; the establishment of research programs within international groups; these activities take place within institutional frameworks that themselves are also multi-scaled.

Especially in climate science, the international level is decisive. This community is characterized by a very strong international structure (which doesn't prevent a diversity of conception and vivid debates), through observational, numerical and institutional global networks and infrastructures—in particular through global research programs like the WCRP or the International Geosphere-Biosphere Programme (IGBP). Sub-programs provided an international framework for field experiments, intercomparison projects and the like, but also a collaborative intellectual dynamic. These different levels are evidently not separated by impenetrable barriers, nor are they independent of each other. The same scientists circulate from the laboratory to the national, European or international level, enter into collaborations, attempt to influence a global research program, drive a project or participate in it.

Research in these fields is oriented by diverse and interdependent logics, in particular by technological advances in computing and observation and by pressing political demands. As this chapter has shown, climate modeling is constituted by diverse cultures of prediction, shaped by different practices, institutional priorities, technical capacities, and modes of domesticating the inherent uncertainties and complexities of modeling the climate system. As such, the future of climate modeling is not mapped out in advance, and although certain tendencies seem to predominate, it remains a matter of debate.

Notes

1   In practice, the equations in the dynamic part include "source terms," which are determined by the physical parameterizations at each time step and for each grid cell.
2   "Climate sensitivity" is defined as the temperature change of the surface of the Earth resulting from a doubling of atmospheric carbon dioxide concentration.

3 Global models rely on "hydrostatic approximation": that is, they disregard the vertical acceleration of air masses and do not calculate convective movements, which are parameterized. High-resolution models take into account vertical pressure variations and calculate convection using the equations of dynamics.

4 GCSS is a sub-program of GEWEX (Global Energy and Water Cycle Experiment), one of the principal research programs of the WCRP.

5 Modeler Christian Jakob compares model developers to an "endangered species." At a WCRP conference in 2011, he asked the audience of more than a thousand researchers to raise their hands if they had changed their models in the previous two years: twenty hands were raised.

6 In this chapter, LMD researchers are identified by a letter followed by a number.

7 White paper on WCRP Grand Challenge. The issue of "Clouds, Circulation and Climate Sensitivity" has been identified by the WCRP as one of the six "Grand Science Challenges" for the next decades. One of its main task is to "tackle the parameterization problem through a better understanding of the interaction between cloud / convective processes and circulation systems."

8 A well-known result of climate model intercomparison exercises is that no model is better than the others (a model can be excellent for certain aspects of the climate and less successful elsewhere) and that the best model is often the mean of all models—which is an argument in favor of a multiplicity of models, rather than convergence toward a single model.

9 The LMD is a laboratory of the Institut Pierre-Simon Laplace (IPSL). The LMD model is the atmospheric component of the IPSL Earth System Model (along with the ocean, biosphere and sea ice components).

10 The Coupled Model Intercomparison Project (CMIP) coordinates worldwide climate model simulations, whose outputs are synthesized in the IPCC Reports. The IPCC's Fifth Assessment Report (AR5), released in the fall of 2013, is based on simulations performed by twenty modeling groups around the world within the CMIP5 framework.

Acknowledgement

The research for this paper was financed by the ANR ClimaConf project.

References

Arakawa, A. (2000) "A personal perspective on the early years of general circulation modeling at UCLA," D.J. Randall (ed.) *General Circulation Model Development: Past, Present and Future*, (International Geophysics Series, vol. 70), San Diego, CA: Academic Press, pp. 2–65.

Bony, S. and J.-L. Dufresne (2005) "Marine boundary layer clouds at the heart of tropical cloud feedback uncertainties in climate models," *Geophysical Research Letters* 32, L20806.

Bony, S., B. Stevens, I. Held, J. F. Mitchell, J.-L. Dufresne, K. A. Emmanuel, P. Friedlingstein, S. Griffies and C. Senior (2011) "Carbon dioxide and climate: perspectives on a scientific assessment," G. Asrar and J. Hurrel (eds.) *Monograph on Climate Science for Serving Society: Research, Modelling and Prediction Priorities*, Dordrecht, Netherlands: Springer, pp. 391–413.

Charney, J. H., A. Arakawa, D. J. Baker et al. (1979) *Carbon Dioxide and Climate: A Scientific Assessment*, Washington, DC: National Academy Press.

Dahan Dalmedico, A. (2010) "Putting the earth system in a numerical box? The evolution from climate modeling toward climate change," *Studies in History and Philosophy of Modern Physics* 41(3), pp. 282–292.

Fleck, L. (1979) *The Genesis and Development of a Scientific Fact*, Chicago, IL: University of Chicago Press.

Galison, P. (1996) "Computer simulations and the trading zone," P. Galison and D. Stump (eds.) *The Disunity of Science: Boundaries, Context and Power*, Stanford, CA: Stanford University Press, pp. 118–157.

GCSS Team (1993) "The GEWEX Cloud System Study (GCSS)," *Bulletin of the American Meteorological Society* 74, pp. 387–399.

Gramelsberger, G. (2010), "Conceiving processes in atmospheric models: General equations, subscale parameterizations, and 'superparameterizations'," *Studies in History and Philosophy of Modern Physics* 41(3), pp. 233–241.

Guillemot, H. (2010) "Connections between climate simulations and observation in climate computer modeling. Scientist's practices and 'bottom-up epistemology' lessons," *Studies in History and Philosophy of Modern Physics* 41(3), pp. 242–252.

Guillemot, H. (2014) "Comprendre le climat pour le prévoir? Sur quelques débats, stratégies et pratiques de climatologues modélisateurs," F. Varenne and M. Silberstein (eds.) *Modéliser et Simuler. Epistémologies et Pratiques de la Modélisation et de la Simulation*, Paris: Editions Matériologiques, pp. 67–99.

Held, I. (2005) "The gap between simulation and understanding in climate modeling," *Bulletin of the American Meteorological Society* 86, pp. 1609–1614.

Hourdin, F., J.-Y. Grandpeix, C. Rio et al. (2013) "LMDZ5B: the atmospheric component of the IPSL climate model with revisited paramerizations for clouds and convection," *Climate Dynamics* 40(9–10), pp. 2193–2222.

Humphreys, P. (2004) *Extending Ourselves: Computational Sciences, Empiricism, and Scientific Method*, Oxford, UK: Oxford University Press.

Jakob, C. (2010) "Accelerating progress in global atmospheric model development through improved parameterizations—Challenges, opportunities and strategies," *Bulletin of the American Meteorological Society* 91, pp. 869–875.

Jakob, C., K. Trenberth, T. Shepherd et al. (2010) *Promoting Model Development*, (White Paper for discussion at the WCRP Modeling Council Meeting, November 2010), Paris, France: WCRP.

Knutti, R. (2010) "The end of model democracy?" *Climatic Change* 102, pp. 395–404.

Lenhard, J. and E. Winsberg (2010) "Holism and Entrenchment and the future of climate model pluralism," *Studies in History and Philosophy of Modern Physics* 41(3), pp. 253–262.

Randall, D., M. Khairoutdinov, A. Arakawa et al. (2003) "Breaking the Cloud Parameterization Deadlock," *Bulletin of the American Meteorological Society* 84, pp. 1547–1564.

Ruphy, S. (2010) "Is the World really 'dappled'? A response to Cartwright's charge against 'cross-wise reduction'," *Philosophy of Science* 70(1), pp. 57–67.

Shackley, S. (2001) "Epistemic lifestyles in climate change modeling," C. Miller and P. Edwards (eds.) *Changing the Atmosphere: Expert Knowledge and Environmental Governance*, Cambridge, MA: MIT Press.

Shukla, J., R. Hagedorn, M. Miller, R. Hagedorn, B. Hoskins, J. Kinter, J. Marotzke, T. N. Palmer and J. Slingo (2009) "Strategies: Revolution in climate prediction is both necessary and possible: A declaration at the world modelling summit for climate prediction," *Bulletin of the American Meteorological Society* 90, pp. 175–178.

Stevens, B. and S. Bony (2013) "What are climate models missing?" *Science* 340(6136), pp. 1053–1054.

Part II

Challenges and debates

Negotiating and using simulation knowledge

8 The (re)emergence of regional climate

Mobile models, regional visions and the government of climate change

Martin Mahony

Introduction

Preceding essays in this volume have documented how various cultures of prediction have combined to make "the climate" something which can be known and, perhaps, governed on a global scale. The understanding of the climate as a dynamic, globally interconnected system has superseded previous understandings of "climate" as a distinctly regional unit, constructed through the spatial delineation of stable weather patterns and types (Heymann 2010; see also Martin-Nielsen in this volume). Since the mid-twentieth century, the mathematical simulation of the climate system has offered a resolutely global picture of complex environmental change, which has been "co-produced" with globalist political ambitions to mitigate the worst consequences of rising temperatures and shifting weather patterns through international political agreements and policies (Miller 2004). However, scientists, campaigners, and politicians have long been aware of the politically paralyzing effects of knowledge claims that refer to abstract, global realities rather than the local realities of everyday existence or routine political decision making. Global climate change, understood through compendia of statistical data, model output and global thresholds, arguably "detaches global fact from local value, projecting a new, totalizing image of the world as it is, without regard for the layered investments that societies have made in worlds as they wish them to be" (Jasanoff 2010: 236).

For Mike Hulme, climates do not travel well across scales. Building a picture of a global climate demands abstraction from the visceral, bodily experience of weather, with all the diverse cultural valences weather has in different places. By "de-culturating" or "purifying climate and letting it travel across scales detached from its cultural anchors," Hulme argues, "we have contributed to conditions that yield psychological dissonance in individuals: the contradictions between what people say about climate change and how they act" (Hulme 2008: 8). We might, therefore, try to understand both the psychological dissonance and political logjam of climate change through the lens of scale— that is, by attending to the construction of particular scales (global, local, or regional) as the most natural, pragmatic, or expedient way of ordering scientific knowledge and political action. This chapter addresses this question directly by exploring an emergent culture of prediction coalescing around Regional Climate Models (RCMs). It is suggested that regional modeling has become a prominent practice of climate science not just because of scientific curiosity or the epistemic drawbacks of global models, but because they function as political technologies of "translation." RCMs have been employed to "spread the news" about climate change—to re-invest the global climate with some of the local meaning of which it is stripped in the moment of its construction. At the same time, RCMs are becoming key tools by which nation-states come to terms with climate change as something that is governable on local scales. Considered as tools of translation, we can understand RCMs as contributing to the processes by which "mobile…associations are established between a variety of agents," linking up "concerns elaborated within rather general and wide-ranging political rationalities with specific programmes for government" in diverse settings (Rose 1999: 50). I will explore the "mobility" or mutability of these associations, in particular through a focus on the mobility of regional models themselves (Callon 1986).

This chapter, while offering a detailed account of the scientific development of regional climate modeling, will therefore also begin an analysis of how this new form of vision has been co-produced with new forms of governmental intervention in relation to climate change. It first traces the emergence of regional modeling in the United States in the late 1980s, before exploring how RCMs have been employed in the production of governmental knowledges in the United Kingdom. These historical snapshots invite questions about the nature of traveling code, or about what happens when models, or bits of models, travel and are re-applied in new settings. The varying mutability of regional modeling tools is explored through a discussion of how RCMs have traveled and been put to work in diverse political contexts, with a particular focus on recent Indian climate politics. Throughout, it is argued that RCMs have participated in the construction of new scales of knowledge and action, have contributed to the mobility of particular forms of climate politics, and are thus an important yet heretofore neglected element in the history of the techniques and practices by which climate change has been rendered governable. RCMs are therefore an important phenomenon through which to understand the development, institutionalization and broader effects of an emergent, distinctive culture of prediction.

The emergence of regional modeling

There are three primary methods by which information on regional-scale climate change can be derived from global climate models. The first is to run a high-resolution atmospheric general circulation model (GCM), which is computationally less demanding than running a coupled ocean-atmosphere model and can therefore feature a smaller model grid. This method has been shown to offer benefits in terms of the simulation of extremes, but the lack of an ocean limits the ability of such models to usefully simulate many climatological phenomena. Related to this method is the employment of a "stretched grid," whereby the resolution of a global model is increased only over a particular region of interest. This technique has been shown to enhance the simulation of phenomena like the West African monsoon in global models (Abiodun et al. 2011). However, the pursuit of regional information has more often been led by two different methods.

"Statistical downscaling" involves the derivation of empirical relationships between global and regional climatic variables, with the former predicting the latter. Once derived, these statistical relationships can be applied to the results of global climate simulations to give results on a regional scale. This technique has been used to study phenomena like hurricanes, river flow, and crop yields under a changing climate (Flato et al. 2013). However, the increasingly more popular method of developing regional climate information, and the one of principal interest here, is known as "dynamical downscaling." This involves the "nesting" of an RCM within a global GCM. The global model provides boundary conditions to the RCM—that is, information about the state of the atmosphere and ocean at the boundaries of the regional domain—as the RCM proceeds through a discrete "time slice" or decadal simulation. Currently, the direction of exchange between the GCM and RCM is mostly one-way, although a few studies have been conducted in which the RCM is able to "feed back" to the GCM, and there is evidence that the improved simulation of small-scale phenomena with an RCM can improve the GCM's simulation of larger-scale phenomena (Flato et al 2013). The following section traces the early history of downscaling simulated global climates to the regional scale.

The search for climatic realism, 1989–1994

The emergence of regional models as tools for simulating the features and futures of regional climates can be traced to the National Center for Atmospheric Research (NCAR) at Boulder, Colorado, in the western United States. In the late 1980s, scientists such as Filippo Giorgi, Gary Bates, and Robert Dickinson began experimenting with the nesting of a meso-scale meteorological model developed at Pennsylvania State University within NCAR's global climate model. However, this early example of traveling code was a case of mutation— the Penn State weather model had to be adjusted before it could be applied to climate studies. The model's physics formulations were either "modified or replaced with more suitable ones," with the most significant alterations being the addition of a full radiative transfer code, as used in the global model, to capture the effects of changing atmospheric compositions, and a scheme for capturing the interactions between soil hydrology and the atmosphere (Giorgi, 1995: 24).

The 1980s was a period of growing scientific and political concern with global climate change. Scientific conferences in the mid-1980s helped diverse scientific communities coalesce around the issue, contributing to the formation of the Intergovernmental Panel on Climate Change (IPCC) in 1988. That same year, NASA's James Hansen testified to the US Congress on an unusually hot summer's day that global warming had well and truly arrived. However, the early efforts of the would-be regional modelers at NCAR were not directed solely at understanding the local and regional effects of global climate change. The late 1980s was also a period when scientists and politicians were grappling with the question of where the United States' growing stocks of spent nuclear fuel could be safely deposited and stored over millennial timescales. While some climatologists were busy exploring the potential environmental consequences of nuclear fallout, others had begun to reckon with the problem of predicting climatic changes—anthropogenic or otherwise—over the vast timescales involved in nuclear waste decision making (Barron 1987). From three prospective sites for the national nuclear dump, Yucca Mountain in Nevada emerged as the preferred option. This isolated ridge within the Great Basin, sandwiched between a nuclear weapons testing site, an Air Force proving ground, and the California border, features a desert climate and the kind of soft, ignimbrite geology which has elsewhere long been exploited for the underground storage of more benign artefacts like wine barrels.

Committing to the storage of nuclear waste at one site for many thousands of years invites questions about the likely evolution of the local climate. However, in the case of Yucca Mountain, GCMs were not deemed adequate tools, in large part due to the influence of topography on the local climate. Yucca Mountain sits in the "rain shadow" of the Sierra Nevada where air moving in from the Pacific rises, cools, and produces precipitation. The area to the west of the Californian mountains therefore receives comparatively little precipitation. This had particular relevance for the nuclear depository debate, as decision makers were keen to know if the groundwater level below Yucca Mountain would ever rise to the depth of the waste stores. The interaction of local climate and local hydrology was therefore of pressing concern, and NCAR received funding from the Department of Energy and from the US Geological Survey's Yucca Mountain Project to launch a research program. A paper written by several NCAR scientists described the inadequacy of a global model for this kind of work:

> The CCM1B [NCAR's GCM] treats the area of Yucca Mountain as though it were on the western slope of the Rocky Mountains, resulting in orographically enhanced rainfall and greatly overestimated initial values of soil moisture.
>
> (Dickinson et al. 1989: 421)

Rather than simulating the Yucca Mountain topography as residing in the lee of the Sierra Nevada, the area was represented by the global model as sitting on the windward side of a mountain chain which elides the Sierra Nevada, and the more easterly Rockies in a low-resolution blob. However, the RCM—or

Limited Area Model (LAM), as it was called at the time—offered greater topographical realism with its 60 km grid (as opposed to the GCM's 500 km grid) and could thus distinguish between the Sierra Nevada and the Rockies, with Yucca Mountain nestled in between. Early experiments with the RCM nested within the GCM, which also featured a biosphere-atmosphere transfer module to simulate evapotranspiration rates, yielded encouraging results for the simulation of Yucca Mountain's local climate. The work reported by Dickinson et al (1989) featured the simulation of the global climate over three years with the GCM, and the production of a January climatology for each year. To explore the "value-added" of the RCM, several "representative" storm systems were simulated, and monthly averages extrapolated. The RCM results were a much better fit to local observations than the GCM output, although the biases of the latter still led to an unrealistic characterization of soil moisture in the former—an issue of particular concern for the nuclear waste plans and their imbrications with the local hydrological cycle (cf. Giorgi and Bates 1989).

In a subsequent paper, Filippo Giorgi reported the use of the RCM to simulate the January climate of the western United States in a more comprehensive fashion, beyond the "simple statistical approach of Dickinson et al." (Giorgi 1990: 942). He again found that the regional simulation offered more realistic averages of temperature and precipitation, which he attributed to the RCM's more realistic topography. Referring to the "long term purpose of applying the model to the generation of climatic change scenarios," Giorgi described plans to explore the performance of the model in different seasons and in different areas of the globe where topography was considered an important driver of regional climate, namely the North American Great Lakes, southeastern Australia and Western Europe (Giorgi 1990: 962).[1]

The development of regional climate modeling thus originated not in a drive to start considering what global climate change might mean on a more local scale, but in a desire to understand the possible future evolution of the climate at a particular point on the map, whatever the drivers of climatic changes might be. Yucca Mountain, or more precisely the few model grid points which corresponded to its geography, became the focus of attention for these regional modelers. However, the possibility of using regional models to characterize the possible impacts of climate change was never far from view.

Soon after the work on the regional climate of the western United States, the NCAR group began publishing studies which explored the differences between GCM and RCM simulations of anthropogenic climate change. During the 1980s, GCMs had been used for equilibrium climate change experiments, whereby the model was run with a baseline atmospheric concentration of carbon dioxide (CO_2), and then with the concentration doubled (cf. Washington and Meehl 1984). Filippo Giorgi and colleagues began using this experimental design to study the differences between GCM and RCM responses. In early publications of this kind, the authors stressed that the results should not be considered as predictions of likely climate change impacts, but as explorations of model response. Of particular interest again, was the issue of topography and its effects on surface climate, and of the effects of sub-GCM grid scale processes on the magnitude and spatial distribution of climatic variables. The RCM was found to generate scenarios with very different spatial structures than the GCM, with differences most pronounced in areas of topographic complexity. Some values, such as precipitation, were found to differ not only in magnitude and spatial structure, but in the very direction of change. These results, the authors suggested, indicated that "cautious use should be made of direct GCM output for impact assessments" (Giorgi et al. 1994: 398).

As this work on the regional climates of the United States was progressing, a scientist from the University of L'Aquila in Italy, where Giorgi had completed his undergraduate and masters studies, visited NCAR to collaborate on the application of the nested modeling technique to studies of western Europe. The visitor, Maria Rosario Marinucci, worked with Giorgi on the validation of the regional model over Europe and then on its use in equilibrium experiments. Similar results were found as for the US runs, again demonstrating the benefits of improved topographical representation and of the ability to simulate climate-hydrology interactions on a smaller scale than before. The western European climate, dominated as it is by complex coastlines and variable topography, emerged as something the future changes of which were best understood through an RCM. Although the biases of the driving GCM still hampered the RCM's realism, especially regarding the simulation of crucial ocean-atmosphere interactions, the authors expressed the opinion "that nested LAM/GCM model systems can provide valuable information on the regional impacts of global climate change" (Giorgi et al. 1992: 10027).

Meanwhile, in the United Kingdom, the hopes for regional climatic realism had also found their way into the center of emerging science-policy relationships, with Shackley and Wynne (1995) suggesting that the greater spatial realism offered by RCMs was attractive to policy makers, as it seemingly offered the "tantalizing prospect" of climatological information directly relevant to local socio-economic processes and resource management. These hopes for regional modeling thus rested on two assumptions of realism: that related to the simulation of local, climatically-relevant topographical details; and that related to a potentially realistic portrayal of local relationships between climate and human societies. As the global climate changed, the hope was that regional models would help make these local relationships governable.

In search of applications, 1995–2002

The period from around 1995 saw a maturation of regional climate modeling as a field of scientific activity. In a confident paper written in 1995, Giorgi referred explicitly to the links between regional modeling and the informational needs of environmental policy-makers. "It is at the regional scale", he argued, "that impacts or, more generally, anthropogenic interactions with the environment are critical for policy-making" (Giorgi 1995: 30). The paper called for new efforts to couple regional climate models—both NCAR's and models starting to emerge at other centers—to models of ecosystems, the oceans and other elements of the "earth system" in order to generate information on, and projections of, human-environment interactions on tight spatial scales. Such work would complement the emerging science of the earth system, pursued largely through highly complex but necessarily coarse-scaled coupled models of diverse atmospheric, hydrological, biological, and physical processes. Giorgi's project would put regional models at the center of a new scientific infrastructure designed to offer scientists and policy makers precise, localized information on the complex environmental interactions which were moving to center stage in international and local politics in the early- to mid-1990s (Jasanoff and Martello, 2004).

Writing later in 2003, several prominent regional climate modelers suggested that RCMs had yet to realize their analytical potential:

> While results from regional model experiments of climate change have been available for about ten years, and regional climate modelers claim use in impacts assessments as one of their important applications, it is only quite recently that scenarios developed using these techniques have actually been applied in a variety of impacts assessments.
>
> (Mearns et al. 2003: 10)

Following a series of what Linda Mearns et al. labelled "pilot studies" concerned with the effects of increased spatial and temporal resolution on climate simulations, around the turn of the millennium, studies started to emerge which used RCM output to study the impacts of climate change on water resources, temperature extremes, agriculture, and forest fires. In many cases, such projects took advantage of the apparent climatic realism offered by RCMs, particularly in regions where topography has a significant impact on mesoscale weather systems. However, few RCM studies in this period systematically analyzed the "value-added" of regional simulation, that is, the different patterns produced in key variables by RCMs as opposed to coarse-resolution GCMs. Meanwhile, the majority of climate change impact studies were simply using the results of global models to drive local-scale impact models, avoiding the use of an RCM altogether (Mearns et al. 2003). However, in the following years the tenor of discussions around the applications of RCMs was set to change, as some actors moved the tools to center stage in discussions of the kinds of knowledge required as societies looked to respond to the challenges of climate change.

Regional prediction as governmental technique, 2002–present

For Linda Mearns and her co-authors of the IPCC guidelines on the development and application of RCM scenarios, the advantages offered by RCMs to impact studies were not self-evident. The benefits brought by regional simulation needed careful consideration, not just through statistical analysis of the differences between GCM and RCM results, but through an initial assessment of the particular needs and resources of a research program. The authors offered a "decision tree" to help analysts decide whether the use of an RCM is really required for their needs, or whether computational and human resources may be better spent on the analysis of global climate scenarios. Considerations such as the ability of GCMs to reliably simulate local climatic features, the variables (e.g. temperature, precipitation, frequency of extremes) of interest, available computational resources, and the spatial scale of analysis, were all factors affecting the choice of whether high resolution information was desirable or necessary. The authors urged researchers to carefully characterize the uncertainties associated with regional simulation and, where appropriate, to offer to policy makers a range of plausible climate change scenarios rather than just a single path through the "cascade" from global climate simulation to local impacts assessment.

As the authors were writing these guidelines, regional climate modeling was moving to the center of governmental efforts to come to terms with possible future changes in climate. In the early 2000s, regional climate scenarios "started to be used to inform, or even to define, statutory public policy guidelines" (Hulme and Dessai 2008: 55) in countries such as the United Kingdom, the United States, and Australia. RCMs, with their ability to offer projections on fine, visually realistic scales, became key tools in this enterprise of making future, local climate change an object of governmental concern and a potential target of governmental intervention.

In 2002, the UK Government published a set of climate projections known as UKCP02. While these were not the first climate projections to be commissioned by the UK Government, they were the first to augment global model results with an RCM. The Hadley Centre's RCM, HadRM3, was connected to an atmospheric model (HadAM3H) and in turn to the Centre's main GCM, HadCM3. A range of IPCC emissions scenarios drove the models, with projections offered for the years 2071–2100 at a scale of 50 km. This was a huge refinement of previous scenarios, which were developed on 300 km or 500 km grids (Hulme and Dessai 2008). Changes were estimated in a number of variables beyond temperature and precipitation, such as cloud cover, diurnal temperature range and relative humidity. Comparisons with the results of other GCM runs enabled the presentation of estimated uncertainty bounds, but the fact that the scenarios themselves were developed using only one family of models (the Hadley Centre models) meant that a fuller quantitative treatment of uncertainty was sacrificed in favor of increased spatial and temporal resolution.

The UK climate scenarios are designed to feed directly into decision making in government, utility provision, and the private sector at all scales. Relevant "stakeholders" have been increasingly drawn into the process of designing the projects with the aim of increasing the usability of the end product. In the case of UKCP02, a panel of prospective users was consulted and expressed a preference for higher spatial and temporal resolutions. Although the subsequent usage of the scenarios was initially limited by the lack of explicit user guidelines, the results found application in a number of sectors, including water and the building sector. Hulme and Dessai (2008) report that actors in the UK water industry found the UKCP02's treatment of uncertainty insufficient for their needs of estimating future supply, whereas the Chartered Institute of Building Survey Engineers drew on the scenarios to revise the heating and cooling design standards recommended to their members.

New scenarios, UKCP09, were published in 2009. This time, an eleven-member "ensemble" of RCMs was used to generate scenarios on a 25 km grid. The ensemble approach enabled the first probabilistic analysis of uncertainty, while the high spatial resolution enabled results to be presented by administrative region, catchment area or, in some cases, marine region. A user interface allowed results to be manipulated by the user, for example to produce maps or graphs of the model output and to explore the uncertainties associated with the different scenarios. Again, a user panel was involved in the discussions of how the results could be delivered in the most useful and usable way.

Tang and Dessai (2012), drawing on interview and survey data of users' experiences with the scenarios, argue that while UKCP09 was seen as highly credible and legitimate science, it arguably fell short in terms of saliency for end users. Although the presentation of Bayesian uncertainty estimates was laudable, it seemingly did not fit easily with the kind of probabilistic information that many decision makers were accustomed to working with. However, Tang and Dessai's analysis of ninety-five adaptation reports, prepared by a number of organizations in accordance with requirements under the Climate Change Act, is testament to the fact that UKCP09 results have been widely used by those concerned with adaptation, to the point where the scenarios have become an "obligatory passage point" (Latour 1987) for those seeking to publicly demonstrate their serious engagement with climate change and their responsible consideration of future risks and uncertainties in their decision making. To this end, the presence of the UK Government's "stamp of approval" has offered legitimacy to the science as a governmental tool, although there are arguably risks associated with huge numbers of actors making decisions based on a single set of visions of the future. However, this model of science and policy, this particular culture of prediction, is being increasingly internationalized, often in the form of mobile climate modeling tools. It is the politics and practices of this mobility to which this chapter now turns.

Models on the move: RCMs as traveling code

While acknowledging that regional simulation may not be necessary for all climate impacts studies, Linda Mearns and colleagues recognized in 2003 that political factors may play a role in the decision to use an RCM at all, and then in the choice of the RCM to be used:

> An important geo-political issue may be the importance of national representation in climate models in the context of international negotiations (i.e., it may matter if a country is or is not on the map).
>
> (Mearns et al. 2003: 21)

Rendering smaller countries visible "on the map" of global climate simulation has been facilitated by a growing family of portable, easy-to-use climate models, offering to users around the world the ability to begin simulating regional climates. For Mearns et al., various factors might influence the circulation and application of these mobile models:

> [V]arious considerations, some of them not strictly scientific, can enter the choice of a given RCM. Among them are model availability, flexibility and user friendliness, consulting support, portability and computing efficiency.
>
> (Mearns et al. 2003: 27)

Although dependent on the outputs of global models, RCMs are computationally lightweight by comparison and thus can often be engineered to run on stand-alone desktop computers and laptops. Freed from the yoke of supercomputers in their vast, air-conditioned halls within the academic and governmental institutions of the global North, the assemblages of code, data, and software which constitute RCMs can now travel relatively easily around the world.

As early as 1995, Filippo Giorgi had begun to consider the potential of RCMs to alter the global geography of knowledge production:

> A possible scenario for future Earth System modeling is that a few large centers can develop global models and provide data which regional centers will use in their regional models to focus on specific issues of relevance for their area.
>
> (Giorgi 1995: 39)

Here Giorgi began to conceptualize regional models as tools which could travel and "be accessed by a broader scientific community" (Giorgi 1995: 39). Around 2003, Giorgi's model, now named RegCM3—an updated version of the model first applied over Yucca Mountain, was mounted on the Internet as a bundle of computer code which could be downloaded and put to use by anyone with the requisite expertise. The model became a central tool for the newly constituted Regional Climate Research Network (RegCNET), based at Giorgi's new institution, the Abdus Salam International Centre for Theoretical Physics (ICTP) in Italy.

Founded in 1964 by Pakistan-born Nobel Laureate, Abdus Salam, the ICTP's mission is to "[f]oster the growth of advanced studies and research in physical and mathematical sciences, especially in support of excellence in developing countries." Describing itself as "a major force in stemming the scientific brain drain from the developing world," the institution supports the training and development of scientists from across the world, and fosters collaborative projects that offer scientists access to new resources and expertise (International Centre for Theoretical Physics 2014).

Following an inaugural meeting in 2003, RegCNET scientists set to work on a variety of collaborative projects, many of which featured work with the new version of RegCM, which had been specifically developed with the simulation of tropical climates in mind (Pal et al. 2007). Publications in a special RegCNET issue of *Theoretical and Applied Climatology* testified to the new international mobility of the RegCM model (Giorgi et al. 2006), and to the capacity of RegCNET scientists to adapt the model to their own climatological requirements and preferences, such as the evaluation and selection of competing forms of cloud parameterization (e.g. Dash et al. 2006).

Comparing Figures 8.1 and 8.2, we can see clearly how the construction of scientific tools involves the engineering of both epistemic and social relations. Such is the case with another itinerant RCM produced by the UK's Met Office Hadley Centre (cf. Mahony and Hulme: 2012). The model—named PRECIS (Producing Regional Climates for Climate Impacts Studies)—was developed in the early 2000s to facilitate the generation of information to support adaptation policy making in developing countries. Unlike RegCM3, which has been promoted more as a tool to conduct climate process studies and to explore climatic variability, PRECIS has always been geared towards policy-relevant applications and the raising of governmental "awareness" of climate change impacts (Jones et al. 2004: 6). PRECIS, as will be explored below, thus arguably extends a local culture of the Hadley Centre, oriented towards what Simon Shackley calls a "hybrid climate modelling policy style," whereby the core objectives and work program of the organization are the outcomes of negotiation with policy makers, rather than the immediate interests of the researchers or model developers (Shackley 2001: 128).

PRECIS consists of the Hadley Centre's regional model packaged within a software program to enable easier operation. The definition of the region to be studied, the boundary conditions and the length of model runs can be selected through a user-friendly interface far removed from the lines of code that underpin the model. PRECIS was designed to be used by those with a high level of knowledge about the climate system, but who perhaps lack direct experience in developing and running climate models. PRECIS has thus migrated beyond the confines of the community of practice defined by what Collins and Evans might call "contributory" expertise in climate modeling, and has spread through a community of practitioners with "interactional" expertise (Collins and Evans 2008).[2] This, at least, is the gradient of expertise constructed along with PRECIS as it has been packaged, tested, and distributed. Initially, a network of "trusted collaborators" was envisaged, who could adapt the model to their local needs, in contrast to those viewed as less capable of autonomously operating the model in a way the Hadley Centre deemed appropriate. In practice, however, the model has largely remained intact on its global travels from the center of UK climate science, maintaining the signifying link between the Hadley Centre label and the Hadley Centre tool inside.

The wide mobility of the PRECIS model has been facilitated in no small part by institutional associations with bodies like the United Nations

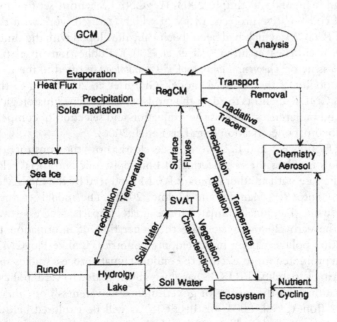

Figure 8.1 Possible design of a regional "Earth System" model.

Source: Giorgi 1995: 38.

Development Programme (UNDP) and the UK Department for International Development (DFID), which have funded training activities and the production of instructional materials because of their own interests in projects to build scientific capacities in developing countries. The model has been specifically promoted as a means for countries to fulfil their requirements under the United Nations Framework Convention on Climate Change (UNFCCC) to develop national adaptation plans based on projected, local climatic changes. PRECIS results have thus appeared in a number of "National Communications" to the UNFCCC, alongside more locally oriented reports and conventional scientific publications, which in turn have been used to bolster political demands like those by some small island states for 1.5 °C to be set as a global warming target, rather than 2 °C (Mahony and Hulme 2012). As an example of "traveling code," PRECIS, therefore, represents an interesting case of how knowledge, data, uncertainties, practices, and assumptions travel together—sometimes in one big bundle, but sometimes in a way which brings different knowledges and assumptions into tension. As a translation device, PRECIS has been positioned as an actor within what has become a dominant and institutionalized approach to adaptation, which positions detailed climate projections at the start of a linear chain of knowledge production and political response.[3] Such an approach arguably produces its own risks and vulnerabilities through the assumption that adaptation is dependent on detailed prediction and thus that adaptation

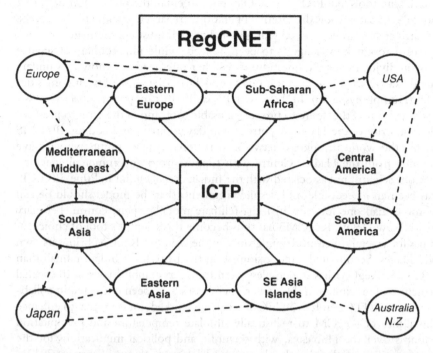

Figure 8.2 A new geography of knowledge production: "Schematic depiction of the structure of the RegCNET."

Source: Giorgi et al. 2006: 2.

strategies can be optimized for a particular predicted future, rather than being robust to a range of possible futures and their inevitable uncertainties (Dessai et al. 2009). In its links with this approach to adaptation PRECIS can thus be considered as part of a "traveling model" in the sense of Behrends et al. (2014)—a suite of mobile technical practices and political assumptions which alter social relations, but which are nonetheless subject to local translations and adaptations (cf. Weisser et al. 2014).

Translation, opacity, and political vision: PRECIS and RegCM in India

Climate scientists in India—specifically, those at the Indian Institute for Tropical Meteorology (IITM) at Pune—have long been among the "trusted collabora-tors" of the Hadley Centre. The collaboration pre-dates PRECIS itself, and Indian scientists have been users of the model since its earliest days. Following PRECIS and its model outputs through the cultural circuits of Indian climate politics is thus a propitious way to explore the politics of regional climate predic-tion and the co-production of "regional" knowledge and a "regional" politics.

Making tools like PRECIS mobile tends to entail the production of "black boxes"—, that is, tools the internal operations are largely opaque to their users. Mearns et al. warned in 2003 that regional models must not be treated as black boxes if robust knowledge is to be produced, while philosophers of science remain divided over the question of whether one needs to know and understand exactly what is going on inside a model to make sense of what comes out of it (Humphreys 2009). In the case of PRECIS, this "epistemic opacity" was a product both of the desire to produce a usable tool, and of the wish to preserve the authority of the Hadley Centre's own development and coding. PRECIS travels the world through a network of national contact points who receive training from the Hadley Centre with support from governmental agencies. For one scientist not associated with the Indian contact point in Pune, PRECIS has been an inaccessible tool despite arguments that the model should be run at more than one location in India (climate modeler, pers. comm, February 2012, New Delhi). RegCM3 has thus become this scientist's tool of choice, as it has for several other modeling groups in the country. RegCM brings its own advantages. Seemingly better at simulating the patterns of Indian rainfall than PRECIS, RegCM has nonetheless been found to exhibit biases in the spatial structure of precipitation. However, such biases have been corrected in collaboration with ITCP in Italy, while others have worked, for example, to enhance the ability of RegCM to realistically simulate temperature and precipitation patterns over the Himalaya, with scientific and political implications for the understanding of glacial melt (climate modeler, pers. comm., February 2012)

Indian climate politics has long been marked by a discourse of scientific and political autonomy. Since climate change emerged on the international agenda in the late 1980s, Indian commentators have been quick to point to historic iniquities of responsibility for climate-warming emissions between developed and developing countries, and to assert the right of the latter to pursue economic growth as the former cut emissions (e.g. Agarwal and Narain 1991). Despite a few modifications along the way, this has been the dominant policy stance of Indian climate negotiators from the birth of the UNFCCC to the present. Alongside this discourse of political autonomy has stood a discourse of scientific autonomy. The charge of "environmental colonialism" against those rich countries who would seek to impinge on India's right to develop has been accompanied by a concern about Western scientists and economists constructing climate change as a problem of equal international responsibility. Outside analyses of India's greenhouse gas emissions have been challenged as over-estimations designed to lead the country down a detrimental political path, while predictions of rapidly melting Himalayan glaciers have been presented by political actors as "alarmist" claims designed to pressure India into renewed political action (Mahony 2014). Meanwhile, scientific knowledge produced in India has been championed as a surer basis on which the country can base its political stance towards international climate negotiations.

The claims and counterclaims about Himalayan glaciers emerged at a time when India's then environment minister, Jairam Ramesh, was gently remolding

India's stance on climate change. During the late 1990s, moves begun to be made to open the country up to market-based forms of emissions reductions, resulting initially in a large number of projects under the UNFCCC Clean Development Mechanism, whereby funding was supplied essentially to offset Northern emissions by developing clean energy projects in the global South. Jairam Ramesh was keen to continue on this trajectory, aligning India with the United States in a preference for a "pledge and review" system of emissions cuts, whereby countries would voluntarily make cuts subject to external monitoring. This move challenged India's long-standing commitment both to refusing binding emissions pledges, and to rendering the Indian economy an object of external monitoring and scrutiny—a challenge to the politico-scientific autonomy described above. To legitimize his political moves, Ramesh thus had to reassert this autonomy through new means. PRECIS stood waiting as a helpful translation device with which to do so.

In advance of the ultimately ill-fated climate talks in Copenhagen in 2009, Ramesh commissioned from a group of government scientists a report on the projected impacts of climate change in India. The group was to coalesce as the Indian Network for Climate Change Assessment (INCCA). With the UKCIP (formerly known as the UK Climate Impacts Programme established in 1997) as an "ideal to follow" (government scientist, pers. comm., March 2012, New Delhi), the report used model output from PRECIS, produced at IITM, to characterize the impacts of climate change in four economically important yet potentially vulnerable regions of the country. Although such studies had been conducted before—not least for India's National Communications to the UNFCCC—the INCCA report was unique in presenting impacts for the 2030s, perhaps in response to the claim in the preceding IPCC report—later found to be erroneous—that Himalayan glaciers could disappear by the year 2035. The INCCA document was championed by Ramesh in the media and the Minister offered a bombastic foreword, which asserted the need for greater "indigenous" scientific input into the formulation of India's political response to climate change:

> We need to make the '3 M's' – Measure, Model and Monitor – the foundation of our decision-making and we need to build indigenous capacity for this. We should not be dependent on external studies to tell us for example about the impact of climate change on our glaciers, on our monsoons, and indeed even on sea level rise.
>
> (INCCA 2010: 9)

Although presented by Ramesh as fundamentally "Indian" science, the INCCA analysis relies on the largely unaltered coding of the British PRECIS model. In this production of governmental knowledge, Ramesh turned to the group of scientists clustered around a nodal point in PRECIS's global network of authorized users, which is dominated by institutions close to national governments. Ramesh did not turn to what some local actors considered a more "Indian" science residing in the network of RegCM users, where local alterations have been made to tailor the model to local circumstances. One RegCM user commented that while the development from scratch of a truly "Indian" model would be a waste of time, a model could nonetheless be considered "ours" after the kind of modifications rendered to RegCM. "Our people have to work hard to call it our model," but starting where others have left off with RegCM was a route to locally reliable and credible science (climate modeler, pers. comm., February 2012, New Delhi).

However, there is little evidence of either RegCM or PRECIS output being used to directly inform adaptation policies. Although the work of PRECIS's Indian users has formed a central pillar of the National Action Plan on Climate Change (NAPCC)—a policy criticized by some as heavily top-down, bureaucratic and inattentive to local socio-ecological realities—the INCCA report was viewed skeptically by those concerned with fostering community-level responses to a changing climate that align needs for adaptation with the demands of community-led development and social justice (environmental campaigner, pers. comm., September 2012, London). PRECIS entered this political fray instead as an actor that could be called upon to legitimize a particular political stance through the provision of trustworthy "local" knowledge. However, to stop the analysis there would be to overemphasize the substantive influence of precisely delineated numerical claims on the formulation of policy. Rather, we can see how PRECIS's regional scenarios were also mobilized to demonstrate and perform an objective engagement with "local" interests. Rather than the application of specific numbers to policy making, the results of PRECIS model runs were used to publicly perform governmental competence in the shifting sands of international climate politics. In the Conclusion, I consider why these public performances are about more than just political rhetoric.

Conclusion: Regional modeling and the government of climate change

Most critical engagements with the use of RCMs in policy-oriented contexts have been content to consider the "legitimacy" of the science itself, with little attention paid to the role of this scientific knowledge in the legitimization of broader forms of politics. From Yucca Mountain in Nevada to the glaciers of the Himalaya, this chapter has traced how regional climate models have offered new fields of "visibility" in an arena—climate change—usually discussed in terms of perfectly global vision. While the "view of everywhere" offered by global models has opened new vistas of planetary interconnectedness, regional models have been employed to explore processes to which global models are largely blind, and to offer to powerful actors a vision of future changes in the local milieu of people and their environments. As such, new knowledge claims have been enabled, concerned with the futures of national territories and the specific vulnerabilities of particular places, and situated within particular assemblages of local and trans-local actors.

This analysis has suggested that RCMs have become one among a suite of techniques by which climate change, as a fundamentally global and post-territorial problem, has been rendered governable by nation-states. RCMs might be considered technologies of translation, linking up "concerns elaborated within rather general and wide-ranging political rationalities with specific programmes for government" in diverse places (Rose 1999: 50). National maps of climate impacts re-territorialize climate change, and enable states to perform a competent engagement with risks and uncertainties that are paradoxically beyond their own capacities of control. Ulrich Beck (1992) labels this the "discursive management of risk," which "is not so much about averting risk but of managing it politically" (Whitehead et al. 2007: 206). While we may dismiss the rhetoric of Jairam Ramesh as political jingoism, such an analysis would overlook the important roles of discourse, display, and performance in the government of climate change, factors which Carl Death (2010) argues are not external to the "real" business of politics, but are directly constitutive of contemporary forms of governmental authority. So, while RegCM has operated as a distinctly mutable mobile tool, being adapted to local climatological circumstances and needs, PRECIS has moved within networks of governmental institutions and thus has been a more prominent agent in performances of governmental authority in relation to climate change.

The emergence, spread, and appropriation—both scientific and political—of regional climate models is an instructive case study in the historical emergence of cultures of prediction. With reference to the aspects of cultures of prediction outlined in Chapter 2, this is a story of how a particular strategy of prediction has always been informed by the perceptions of scientists and their backers of a particular social role for predictive knowledge—in the case of RCMs, the technical and discursive management of risk. This has been enabled by a set of computational practices which prioritize spatial realism as a means of securing both reliable results and an ability to make "visible" the effects of climate change to interested local parties, such as local government decision-makers, in a way that alternative practices, such as statistical downscaling, cannot. Nonetheless, the practices of regional modeling, and the material forms through which such practices have been made globally mobile, give rise to significant challenges related to uncertainty. As argued in Chapter 2, domesticating uncertainty is as much a social as an epistemic process, and in the case of mobile RCMs it is a process shaped by the design of software interfaces which 'hide' a model's core code (e.g. Lahsen 2005), by considerations of who is sufficiently expert to be permitted to access and change that code, by the material realities of scientists in resource-poor (and thus computationally limited) institutions being unable to stabilize predictive claims by making multiple iterations of them, and, perhaps, by the different institutional locations of scientists and the broader politics of managing uncertainty at the science-policy interface (see Barnes 2016).

As the comparison of PRECIS and RegCM shows, it is important to attend to how the details of different techniques and tools enable different kinds of social relations and practices to come into being, as evidenced by the very different institutional networks through which these models move. Nonetheless, it is an overall contention of this chapter that the global circulation of regional climate modeling tools is a key process through which we can comprehend the broader epistemic, social and political effects of contemporary cultures of prediction in the science and politics of environmental change.

Notes

1 The Yucca Mountain proposals were approved in 2002 but then shelved in 2011 amid much political controversy.
2 Contributory expertise is defined as that which belongs to those who contribute directly to an area of scientific activity, like model developers. Interactional experts are those who are conversant in a field or practice, but who do not necessarily seek to advance its frontiers.
3 Dessai et al. (2005) label this "the IPCC approach" (see also Hulme 2011).

References

Abiodun, B. J., W. J. Gutowski, A. M. Abatan and J. M. Prusa (2011) "CAM-EULAG: A non-hydrostatic atmospheric climate model with grid stretching," *Acta Geophysica* 59(6), pp. 1158–1167.

Agarwal, A. and S. Narain (1991) *Global Warming in an Unequal World: A Case of Environmental Colonialism*, New Delhi, India: Centre for Science and Environment.

Barnes, J. (2016) "Uncertainty in the signal: modelling Egypt's water futures," *Journal of the Royal Anthropological Institute* 22(S1), pp. 46–66.

Barron, E. J. (1987) "Nuclear waste disposal: a climate problem?—An editorial," *Climatic Change* 10, pp. 107–109.

Beck, U. (1992) *Risk Society: Towards A New Modernity*, London: Sage.

Behrends, A., S.-J. Park and R. Rottenburg (eds. 2014) *Travelling Models in African Conflict Management: Translating Technologies of Social Ordering*, Leiden, Netherlands: Brill.

Callon, M. (1986) "Some elements of a sociology of translation: Domestication of the scallops and the fishermen of St Brieuc Bay," J. Law (ed.) *Power, Action and Belief: A New Sociology of Knowledge?* London: Routledge & Kegan Paul, pp. 196–233.

Collins, H. and R. Evans (2008) *Rethinking Expertise*, Chicago, IL: University of Chicago Press.

Dash, S. K., M. S. Shekhar and G. P. Singh (2006) "Simulation of Indian summer monsoon circulation and rainfall using RegCM3," *Theoretical and Applied Climatology* 86(1–4), pp. 161–172.

Death, C. (2010) *Governing Sustainable Development: Partnerships, Protests and Power at the World Summit*, London: Taylor & Francis.

Dessai, S., X. Lu and J. Risbey (2005) "On the role of climate scenarios for adaptation planning," *Global Environmental Change* 15(2), pp. 87–97.

Dessai, S., M. Hulme, R. Lempert and R. Pielke Jr. (2009) "Climate prediction: a limit to adaptation?," W. N. Adger, I. Lorenzoni and K. O'Brien (eds.) *Adapting to Climate Change: Thresholds, Values, Governance*, Cambridge, MA: Cambridge University Press, pp. 64–78.

Dickinson, R. E., R. Errico, F. Giorgi and G. T. Bates (1989) "A regional climate model for the western United States," *Climatic Change* 15, pp. 383–422.

Flato, G., J. Marotzke, B. Abiodun et al. (2013) "Evaluation of climate models," T. F. Stocker, D. Qin, G.-K. Plattner et al. (eds.) *Climate Change 2013: The Physical Science*

Basis. Contribution of Working Group I to the Fifth Assessment Report of the Intergovernmental Panel on Climate Change. Cambridge, MA: Cambridge University Press pp. 741–866.

Giorgi, F. (1995) "Perspectives for regional earth system modeling," *Global and Planetary Change* 10(94), pp. 23–42.

Giorgi, F. (1990) "Simulation of regional climate using a limited area model nested in a general circulation model," *Journal of Climate* 3, pp. 941–963.

Giorgi, F. and G. T. Bates, (1989) "The climatological skill of a regional model over complex terrain," *Monthly Weather Review* 117, pp. 2325–2347.

Giorgi, F., M. R. Marinucci and G. Visconti (1992) "A 2XCO2 climate change scenario over Europe generated using a limited area model nested in a general circulation model 2. Climate change scenario," *Journal of Geophysical Research* 97, pp. 10011–10028.

Giorgi, F., C. S. Brodeur and G. T. Bates (1994) "Regional climate change scenarios over the United States produced with a nested regional climate model," *Journal of Climate* 7, pp. 375–399.

Giorgi, F., J. Pal X. Bi, L. Sloan, N. Elguindi and F. Solmon (2006) "Introduction to the TAC special issue: The RegCNET network," *Theoretical and Applied Climatology* 86, pp. 1–4.

Heymann, M. (2010) "The evolution of climate ideas and knowledge," *Wiley Interdisciplinary Reviews: Climate Change* 1(4), pp. 581–597.

Hulme, M. (2008) "Geographical work at the boundaries of climate change," *Transactions of the Institute of British Geographers* 33(1), pp. 5–11.

Hulme, M. (2011) "Reducing the future to climate: A story of climate determinism and reductionism," *Osiris* 26(1), pp. 245–266.

Hulme, M. and S. Dessai (2008) "Negotiating future climates for public policy: a critical assessment of the development of climate scenarios for the UK," *Environmental Science & Policy* 11(1), pp. 54–70.

Humphreys, P. (2009) "The philosophical novelty of computer simulation methods," *Synthese* 169(3), pp. 615–626.

INCCA Indian Network for Climate Change (2010) *Climate Change and India: A 4x4 Assessment*, New Delhi, India: Ministry of Environment and Forests.

ICTP International Centre for Theoretical Physics (2014) *Homepage*, available from: <http://www.ictp.it/>, accessed 20 August 2014.

Jasanoff, S. (2010) "A new climate for society," *Theory, Culture & Society* 27(2–3), pp.233–253.

Jasanoff, S. and M. L. Martello (2004) *Earthly Politics: Local and Global in Environmental Governance*, Cambridge, MA: MIT Press.

Jones, R., M. Noguer, D. Hassell et al. (2004) *Generating High Resolution Climate Change Scenarios using PRECIS*, Exeter, UK: Met Office Hadley Centre.

Lahsen, M. (2005) "Seductive simulations? Uncertainty distribution around climate models," *Social Studies of Science* 35(6), 895–922.

Latour, B. (1987) *Science in Action: How to Follow Scientists and Engineers through Society*, Cambridge, MA: Harvard University Press.

Mahony, M. (2014) "The predictive state: Science, territory and the future of the Indian climate," *Social Studies of Science* 44(1), pp. 109–133.

Mahony, M. and M. Hulme (2012) "Model migrations: mobility and boundary crossings in regional climate prediction," *Transactions of the Institute of British Geographers* 37(2), pp. 197–211.

Mearns, L. O., F. Giorgi, B. Whetton, D. Pabon, M. Hulme and M. Lal (2003) *Guidelines for Use of Climate Scenarios Developed from Regional Climate Model Experiments*, Geneva, Switzerland: IPCC Intergovernmental Panel on Climate Change.

Miller, C.A. (2004) "Climate science and the making of a global political order," S. Jasanoff (ed.) *States of Knowledge: The Co-Production of Science and Social Order*, London: Routledge, pp. 46–66.

Pal, J. S., F. Giorgi, B. Bi et al. (2007) "Regional Climate Modeling for the Developing World: The ICTP RegCM3 and RegCNET," *Bulletin of the American Meteorological Society* 88(9), pp. 1395–1409.

Rose, N. (1999) *Powers of Freedom: Reframing Political Thought*, Cambridge, MA: Cambridge University Press.

Shackley, S. (2001) "Epistemic lifestyles in climate change modelling," C. A. Miller and P. N. Edwards (eds.) *Changing the Atmosphere*, Cambridge, MA: MIT Press, pp. 107–134.

Tang, S. and S. Dessai (2012) "Usable science? The UK climate projections 2009 and decision support for adaptation planning," *Weather, Climate, and Society* 44, pp. 1–45.

Washington, W. M. and G. A. Meehl (1984) "Seasonal cycle experiment on the climate sensitivity due to a doubling of CO_2 with an atmospheric general circulation model coupled to a simple mixed-layer ocean model," *Journal of Geophysical Research* 89(D6), p. 9475.

Weisser, F., M. Bollig, M. Doevenspeck and D. Müller-Mahn (2014) "Translating the 'adaptation to climate change' paradigm: the politics of a travelling idea in Africa," *The Geographical Journal* 180(2), pp. 111–119.

Whitehead, M., R. Jones and M. Jones (2007) *The Nature of the State: Excavating the Political Ecologies of the Modern State*, Oxford, UK: Oxford University Press.

9 Bellwether, exceptionalism, and other tropes

Political coproduction of Arctic climate modeling

Nina Wormbs, Ralf Döscher, Annika E. Nilsson and Sverker Sörlin

Introduction

There are many processes that open our eyes to the possible effects of climate change. Those occurring in the Arctic are among the most striking for at least two reasons. One is because their mediation plays on feelings for species that have gained iconic status, and the other is that the impact of climate change has become visible in ways that are easily understood even by laypeople. The 2007 sea ice minimum is one such example (Christensen et al. 2013). It can be seen as an early outcome of the real-time experiment of global climate change and has contributed to giving the Arctic a particular place in the popular understanding of its impacts.

The idea that the Arctic has a special role in global climate change is far from new. Early research on the dynamics of ice ages already focused on the Arctic in the late 1800s (Crawford 1998; Sörlin 2009), and even in the 1930s ideas were formulated about climate-relevant processes being reinforced in the far north (Ahlmann 1936, 1943; Wright 1953; Sörlin and Lajus 2013). This idea—later labeled "Arctic amplification"—was rearticulated in the 1980s based on insights from computer-based climate science, which again put the spotlight on the region.

The modeling of Arctic climate change is much more recent and started only in the 1990s. Given the long history of interest in Arctic climate change, this late introduction of the Arctic to climate modeling is somewhat of a paradox. The central theme of this article is to address the issue of why the Arctic arrived so late to climate modeling. It also aims to contribute to a fuller picture of how, when, and why the Arctic started to play a central role in climate discourse and for the understanding of global climate change. The analysis shows the close interconnections between scientific motives and political context and the need to combine the two in our understandings of new cultures of prediction. The article thus presents a case for understanding the "science politics of climate change" as a single, integrated discourse with strong feedbacks between its different threads.

The theoretical foundations lie in the understanding that science and politics are co-produced (Jasanoff 2004; Jasanoff and Wynne 1998; Miller 2004) and that scientific developments cannot be explained without including driving forces external to the science (Kuhn 1962). By providing an integrative and contextual narrative, the article aims to provide a complement to accounts that focus mainly on the scientific processes (Weart 2008).

The analysis highlights the appearance of some prominent tropes of the Arctic region. One is a long discursive tradition of Arctic exceptionalism, that the Arctic is unique and separate from the rest of the world (Bravo 2009). How then should we understand why and how the Arctic has come to be viewed as a bellwether for "global" change? Another trope is the Arctic as a laboratory and field site for exploration by and for the benefit of national interests largely located outside the region (Doel et al. 2014). This is a trope that has been challenged more recently, not least by political developments in the circumpolar north. How can the relationship between this shift and the science of climate modeling be understood?

This paper joins a growing corpus of scholarly work in recent years calling for greater appreciation for the specificity of different climate discourses (Ross 1991; Hume 2009: 325, 330; Endfield 2011). Climate and climate change have become global categories, predominantly represented by a single scientific narrative, that has obscured both climatic and political distinctions between regions, places, and timescales and which also runs the risk of disregarding the capacities of specific cultures and societies to understand and adapt to climate change and its consequences. Understanding climatic change means returning to these particularities, as they are manifest for example in distinctive cultures of prediction, in an attempt to reassemble them into a nuanced picture of the whole.

Arctic climate change before models

Insights about the importance of the Arctic climate to regions beyond the Arctic date back to the late 1800s. Svante Arrhenius's analysis of the greenhouse effect was linked to an attempt to understand the causes of ice ages (Crawford 1998). Warm periods have been reported at various times in history and have been scientifically documented for the Barents Sea region since at least circa 1920 (Knipowitsch 1921), along with observations of diminishing Arctic sea ice (Zubov 1933; Lajus and Sörlin 2014). Similar results emerged from the growing science of glaciology in the 1920s and the 1930s (Sörlin 2009). Glaciological work in many countries gradually showed a pattern of diminishing glaciers, a process that seemed to start around 1900 and continued at least well into the 1940s. Temporary glacier reduction had been part of local knowledge in different parts of the world, including the Arctic, since time immemorial (Cruikshank 2005; Carey 2010; Krupnik et al. 2010), but was never conceived of as part of any systematic pattern before the modern ice age theory arrived around 1840 (Macdougall 2004).

Up to the early decades of the twentieth century, whether glaciers were growing or shrinking was established through manual observation or repeat photography. As it soon became obvious that long-term trends in glacier behavior could be caused by changes in climate, Swedish geographer Hans Ahlmann's idea of "polar warming" gained traction in the 1940s and early

1950s (Sörlin 2011). However, this image of polar warming was not connected to the meteorological sciences. Rather, it represented a scientific culture quite different from what later developed into climate modeling (Heymann 2009, 2010; Sörlin 2009, 2011).

By contrast climate modeling has its roots in the new school of dynamic meteorology that developed in response to the need to better understand and predict weather (Harper 2008). Meteorology requires networks of observations, so meteorologists started early to collaborate across national borders (Crawford 1992). During the war years, military needs pushed this and other earth sciences further, including attention to the Arctic—but not, of course, in an international collaborative context (Doel 2003). However, once World War II was over, internationalism was fostered and institutions were created that succeeded in bringing scientists together across the East-West divide. These included the World Meteorological Organization (WMO), created as a special agency under the United Nations in 1951 (Miller and Edwards 2001). Since the 1960s, the WMO has been one of the major international platforms for cooperation among climate scientists.

One significant event was the International Geophysical Year (IGY) 1957/58 (Launius et al. 2010; Krupnik et al. 2009). It not only saw the launch of the first satellites, but also cooperation in data collection (Nilsson 2009a). The IGY and the preparations for it became the starting point for serious Antarctic research as well as for Antarctic political cooperation, epitomized in the Antarctic Treaty 1959 (Elzinga 1993; Shadian and Tennberg 2009). The Cold War space race contributed to improved opportunities for meteorologists and geophysicists to gather data about the weather conditions high in the atmosphere. Also crucial was the first generation of computer models for weather prediction (Harper 2008). Combined, these technologies made the collection and processing of data possible in remote areas (Edwards 2004a, 2004b, 2010).

In the Arctic, Cold War tensions ran high, not least due to the strategic role of the region for transcontinental nuclear missiles. The highly securitized (Heininen 2004; Åtland 2008; Paglia 2016) political atmosphere in the Arctic obviously put serious constraints on data gathering and scientific collaboration. As late as the 1980s, the Arctic was still heavily militarized and access to study areas was a major concern; in addition, the United Nations Convention on the Law of the Sea (UNCLOS) and new zones of sovereignty made areas that had been open only a decade earlier inaccessible to research, for example, hampering research on the Soviet Union's northern continental shelf (Johnson Theutenberg 1982). This setting for research collaboration can be placed in contrast with Antarctica, where the IGY had led to an international institution to coordinate research: The Scientific Committee on Antarctic Research (SCAR).

The Arctic in early global climate models

Despite the fact that the Arctic had held a position in the thinking on climate change both in the region and globally for more than a century, the region did not play a prominent role in the emerging computer modeling of the climate, and neither did polar research. In 1967 the first global ocean climate model was published by Bryan and Cox (1967); around the same time, a model combining ocean and atmosphere components, the first "coupled" model, was explored by Manabe and Bryan (1969). It did not yet feature a realistic geography, however. Among other things, both polar regions were cut off, because they were difficult to handle technically given the convergence of longitudes toward the poles (Griffies et al. 2000). In much the same way, the poles were mercilessly dismissed from the global warming projection that was performed by Manabe and Wetherald (1975). It was based on an atmosphere-stand-alone model by Manabe (1969) and a simple description of the ocean surface. The model domain was limited, cutting off polar areas, ocean more than land. Land areas were cut off north of 81.7°N, ocean areas north of 66.5°N.

The first Atmosphere–Ocean General Circulation Model (AOGCM) to include a complete latitudinal representation of polar oceans and land was documented in a pair of papers by Manabe, Bryan, and Spelman (1975) and Bryan, Manabe, and Pacanowski (1975). That model setup represented, for the first time, a physical ocean–atmosphere core of the climate system in a geography like the real world, albeit coarsely. A simple sea ice model was implemented as well, with the ice that could move and distort in addition to melting and freezing. The area of Arctic sea ice coverage was simulated in a realistic way, although the thickness of the ice was exaggerated. This was caused by the model's failure to account for the export of ice southwards into the Greenland Sea, which led to accumulation of ice volume in the simulation.

Despite this unrealistic representation of the Arctic, when the Manabe and Wetherald (1975) model simulations showed greater warming in the remaining high latitudes than in the tropics, these early climate models brought attention to a notion that the Arctic had a special role to play for climate change. Broecker (1975) introduced the term "polar amplification," referring to the greater warming in the Arctic region compared to the Northern Hemisphere or to externally forced global warming (Broecker 1987; Dansgaard et al. 1982).

The amplification was further identified in a global circulation model (GCM) study by Manabe and Stouffer (1980). Manabe and Wetherald (1975) attributed the high-latitude amplification of the global change signal to what is known as "ice-albedo feedback." This designates a self-amplifying process by which land ice and snow cover are reduced due to atmospheric warming, followed by a reduction of surface albedo (ability to reflect solar radiation). Consequentially, the absorption of heat by the surface is increased, leading to an even warmer surface, which eventually warms even the overlying atmosphere. The ice-albedo feedback was not unexpected in principle; in fact, Arrhenius had already pointed to it in his ambition to explain ice ages (Weart 2008; Serreze and Francis 2006, 2006a). By the late 1960s, its importance for the mean earth temperature had been elaborated by Budyko (1969). However, to provide a realistic representation of its effects on sea ice and the ocean in climate models, there is also a need to allow for differences between the albedo of snow (high), ice (moderate), and open water (low) to represent sea ice cover. Rendering ice thickness is also required to enable useful studies of the effects on the overall extent of sea ice. This level of data was not generally incorporated into the first general circulation models (GCMs), for example, the Manabe and Stouffer (1980) model, as the evolution of Manabe, Bryan and Spelman (1975) did not incorporate different albedos for snow and ice (Dickinson 1986: 225). As a consequence, the ice-albedo feedback did not affect the model simulation until all ice and snow in a given grid element had been replaced by water. Later, more sophisticated albedo formulations took parameters such as ice thickness, ice age, and snow cover and their effect on albedo into account in a way that provides an earlier, more realistic start of the feedback process.

Even if the climate models of the 1970s had many known weaknesses, scientists' own confidence in the models had grown. This was evident, for example, in the 1979 report from the National Academy of Sciences in the United States, known as the "Charney Report" after the chairperson of the committee that authored it, mathematician Jule Charney, that established global warming as real. The scientists' warnings met with increased interest by the media and in the policy sphere, just as earlier warming theories had met with media and policy interest, including from the military (Sörlin 2009, 2011; Weart 2008). Earlier warming news was commonly greeted with enthusiasm; from the 1970s on it was rather framed as an international environmental concern as attention had shifted to include negative impacts, along with concerns, arising from the discussion on ozone depletion in the 1980s, that human activities could indeed influence the Earth's atmosphere (Litfin 1994). Important events in the reframing of opinion were the 1972 UN Conference on Human Environment in Stockholm, the First International Climate Conference in 1979 (Weart 2008), the creation of the World Climate Research Program in 1980 (Agrawala 1998a), and a series of workshops that were held in Villach, Austria, which led to an international assessment of the impact of climate change that was published in 1986 (Bolin et al. 1986; Franz 1997). The message that emerged made a clear connection between the scientific understanding of climate change and the need for policy action (Bolin et al. 1986). In short, climate change research, based on developments in meteorological modeling, was becoming established and was on its way toward institutionalization.

The Arctic did not play a prominent role in the 1986 report, but the authors included statements about Arctic amplification in their discussion about model performance: "simple climate models show that the albedo decreases from reduction in the cover of snow and ice can significantly amplify the climate warming of increased CO_2" (Bolin et al. 1986: 224). They also emphasized many model deficiencies such as the unrealistic treatment of albedo, unsatisfactory models of cloud properties at the sea-ice margin, oversimplified dynamics of the sea ice itself, and the inconsistency of the numbers for albedo used in the models with available observational studies.

Other contemporary reports also showed that Arctic amplification was receiving scientific attention in the 1980s, including observations that the model representation of the amplification depended on details of the sea ice model used to motivate field experiments to improve knowledge about sea ice processes to support model development (National Research Council 1986). Another branch of the discussion was focusing on possible effects of amplification on meridional ocean circulation (Schlesinger 1988). An Arctic amplification of global warming would mean a reduced temperature gradient between polar and subtropical latitudes. This could be a driving force for reduced thermohaline driving of the global ocean circulation and also explain observed proxy-record behavior (Genthon et al. 1987).

The birth of "climate politics"

By the mid to late 1980s the interest in Arctic climate change experienced a major shift. The general background was a vast and comprehensive introduction of environmental concerns not only in a range of individual nations, but also on the global level with the 1987 UN report "Our Common Future", and in the run-up to the Rio global summit in 1992. More specifically it was a rapid shift in the policy framework for climate change science, in essence the emergence of climate politics, with the formation of the Intergovernmental Panel on Climate Change (IPCC) (Agrawala 1998a, 1998b) and the signing of the UN Framework Convention on Climate Change (UNFCCC). The formation of this global climate governance regime in the period from 1986 to 1992 coincided with the end of the Cold War. What role did the Arctic play in this regime shift?

The Arctic had no significant part in the creation of a global framework for climate governance. It was mainly recognized by the scientists working with the models as described earlier, and the political discourse was more focused on the relationship between developed and developing countries. However, the creation of the UNFCCC and demands on cutting emissions of greenhouse gases raised the political stakes related to climate science. As the UNFCCC had been formulated it became increasingly important not only to provide indisputable evidence of climate change as such, but also to prove that it was caused by anthropogenic emissions. This political context led to stronger political demand for climate change data and in particular for subjecting research on models and scenarios to more scientific quality control. Climate scientists knew that the Arctic was important, not only for further data for the models, but also as a region where climate change was likely to be observed. Indeed, the demand for more knowledge about climate processes in this data scarce region served as a major incentive in the creation of the International Arctic Science Committee (IASC) in 1990 (Nilsson 2007). The first IPCC report (1990) mentioned that the polar regions could warm two to three times more than the global mean and that some models predicted an ice-free Arctic with profound consequences for marine ecosystems. In addition to the ice-albedo feedback, it also discussed other potential feedback mechanisms such as the Arctic tundra and boreal wetlands as sources of methane and the role of the

North Atlantic for the mixing of ocean water and heat transport (Houghton et al. 1990).

Even if the need for more climate related knowledge about the Arctic was well recognized, it was no straightforward task to get to the Arctic to perform the research needed. A major change in that regard came with the end of the Cold War and the transformation of the region from a secret zone of potential missile warfare to one of multinational interest with sensitive environmental properties.

The most pronounced expression of this policy change was President Gorbachev's 1987 Murmansk speech on the Arctic, which he portrayed as a "zone of peace," highlighting the need for scientific and environmental coop-eration. This was seen as a precondition for being able to develop the vast resources of the region that had come increasingly into focus with the devel-opment of offshore technology to explore and exploit hydrocarbon deposits under the continental shelf.

This strong interest in desecuritizing the Arctic (Åtland 2008) was a new political context that created opportunities for climate science. A group of scientists who had been active in the Antarctic scientific cooperation started informal discussions in 1986 and concluded that it was time for Arctic countries to start a circumpolar joint scientific effort (Nilsson 2007). Three scientists later prepared a report that laid the groundwork for formal cooperation in IASC, the main aim of which was to increase knowledge about Arctic processes, with an emphasis on the global significance of changes in Arctic climate, weather, and ocean circulation (Archer and Scrivener 2000).

This development can be seen as a re-emergence of one of the old tropes: the Arctic as a resource and laboratory for actors with their primary interests elsewhere. But there was also a shift. The Arctic was now placed into a larger global context that had not been part of the earlier exceptionalism discourse. A new discourse was taking over, based on an understanding of the climate systems as being global in character.

Parallel to the development of scientific cooperation, an explicitly politi-cal process resulted in formalized circumpolar cooperation in the Arctic Environmental Protection Strategy (AEPS) (Young 1998; Tennberg 2000). While climate science remained a central part of IASC's activities, it initially played a limited role in emerging political cooperation, as climate change was defined as a global issue rather than an Arctic concern and that responsibility for such, therefore, was to be placed elsewhere, for example, with the IPCC. The mandate given to the Arctic Monitoring and Assessment Programme (AMAP), which was the relevant working group within the political coopera-tion, therefore, did not include knowledge needs directly relevant to modeling. It was instead "to review the integrated results of these programs with the view of identifying gaps in the scope of the monitoring and research under these fora and with a view to ensuring that specific issues related to the Arctic region are placed on the agenda of appropriate international bodies" (Arctic

Environmental Protection Strategy 1993). AMAP's first major scientific assessment included a chapter on climate change, ozone, and ultraviolet radiation and concluded that "it is clear that state-of-the-art atmospheric models do not adequately simulate the present Arctic climate" (Weatherhead 1998: 729).

By 1999, AMAP had started to prepare for a more focused assessment of climate change. Meanwhile, personal links between the IPCC and IASC were coming into the picture involving IPCC's chair at the time, Bert Bolin, and Robert W. Corell, global change researcher and US representative and chair of IASC's regional board at the time. Corell and Bolin had identified the need to complement the global scale focus with more regional scale studies, and their choice fell on looking more closely at the Arctic. What made the Arctic attractive was that there was a sufficient amount of research to know that the signals of Arctic warming were very strong, and that IASC and the Arctic Council offered an organizational setting in which such an assessment could be carried out (Corell, interview 2004).[1]

The two processes (AMAP and IPCC/IASC) were eventually merged and became the starting point for the Arctic Climate Impact Assessment (ACIA), which was formally launched in 2000, although regional Arctic models were excluded from the first ACIA because they were not sufficiently mature to run coupled regional scenario projections (ACIA 2005; Nilsson 2007). Regional climate scenario efforts had been carried out for Europe at that time (e.g. the Swedish SWECLIM project 1996–2003), but the coupled model tools were not developed enough to work satisfactorily for a more complex Arctic domain.

The developments described here show that the changing political climate in the Arctic and the new cooperative environment that had started to emerge during the 1990s created a new environment for research in the Arctic, and also for making connections between the scientific and policy spheres regarding Arctic climate change. They coincided with the rise of several new research programs that were aimed specifically at trying to understand Arctic processes as a way to improve climate models. Among these was the Surface Heat Budget of the Arctic Ocean experiment (SHEBA) in 1997/1998, which was motivated by the large discrepancies among simulations by GCMs of the present and future climate in the Arctic, and by uncertainty about the impact of the Arctic on global climate change. The US-led Study of Arctic Environmental Change program (SEARCH) began in the mid-1990s after a number of scientists had become concerned about the magnitude of the changes they were observing in Arctic Ocean and atmospheric conditions (ARCUS, 2014). A third major effort was the Arctic Climate System Study (ACSYS), which was a project within World Climate Research Programme (WCRP) aimed at understanding the role of the Arctic in the global climate and improving the representation of the Arctic region in coupled atmosphere-ice-ocean models. All of these programs were greatly facilitated by the fact that the Arctic was no longer caught in Cold War international relations.

The sea ice component

Sea ice models require observational data covering the state of ice and related processes. Until recently, such data were scarce, especially for ice thickness. There had been measurements by vessels traveling the seas, by annual ice camps and short-term observation campaigns on the ice (Lajus and Sörlin 2014). However those observations were difficult to compare and to integrate into a complete picture. Additional measurements had been performed with military submarines since the 1950s, but complete data sets were not released until the mid-1990s (Rothrock et al. 1999). All of these data were difficult to access and did not comprise very long series (Houghton et al. 1990: 224, 225).

Until the 1990s, sea ice was largely accounted for by vertical thermal processes connected to freezing and melting. In the second IPCC Assessment Report, published in 1995, a link was made between better knowledge about sea ice dynamics and climate sensitivity − a problem fundamentally connected with the credibility of the GCMs and the whole idea of simulating substantial global warming as a response to emissions of greenhouse gases. The link to sea ice was made based on experiments done by different groups, in which the physical features of ice had been accounted for through parameterization (Gates et al. 1995: 269; see also Guillemot in this volume). The dynamics of ice, such as how it reacts to stress, strain or skewing, known as rheology, can be incorporated differently and more or less comprehensively into models. Among the first attempts to design a sea ice model with dynamic rheology was Hibler's (1979). This approach took into account movement, deformation and freezing/melting processes, but was not actually incorporated into a few GCMs until the 1990s (Bitz et al. 2012).

The first GCMs were developed by individual groups without systematic interaction. During the early 1990s, this approach was supplemented by comparing different GCMs. Specific Model Intercomparison Projects (MIPs) were initiated within meteorology and oceanography (Sundberg 2011). In 1995 the WCRP launched the Coupled Model Intercomparison Project (CMIP) as a standard experimental protocol for studying the output of coupled AOGCMs, and as a way for a diverse community of scientists to analyze GCMs in a systematic fashion (CMIP 2014). Virtually the entire climate modelling community participated; it has been described as a sign of membership in the community of global climate modelers (Sundberg 2011: 9).

In the case of scenario projections, GCMs also provide large scale input to regional climate models (RCMs). Comparisons between RCMs and GCMs raise new questions on where improvement should be made, and RCMs provide a more detailed interpretation of global scenario projections (Rummukainen 2010; see also Mahony in this volume). Regional intercomparison projects more firmly focused on understanding Arctic processes included the Sea Ice Model Intercomparison Project (SIMIP), the Arctic Regional Climate Intercomparison Project (ARCMIP) and the Arctic Ocean Model Intercomparison Project (AOMIP).

SIMIP was carried out in the second half of the 1990s as part of ACSYS, as a way to better understand and improve the representation of sea ice in global climate models. SIMIP, as the other Arctic model intercomparison projects (MIPs), used observation data collected within the SHEBA experiment in 1997/1998. Still, despite advancements in Arctic climate science in the late 1990s, data on Arctic conditions were not sufficiently up to date when the third IPCC assessment report was published in 2001. In the discussion of the sea ice component of models, the IPCC authors concluded that there was "a slow adoption, within coupled climate models, of advances in stand-alone sea ice and coupled sea-ice/ocean models" (McAveney 2001: Chapter 8.5.3). Several broad questions were identified for ARCMIP. One concerned identifying which processes were inadequately represented by regional models, and another focused on the level of certainty. The project was also to investigate how regional models could best be used for scenarios. Arctic RCMs would help understand whether simulation problems and climate variability were of Arctic origin or rather brought in from lower latitudes. The Arctic Ocean Model Intercomparison Project (AOMIP) began in 2001 as an effort to identify systematic errors in Arctic Ocean models and to reduce uncertainties in climate projections, raising demands for improved descriptions of Arctic Ocean circulation and sea ice conditions (Proshutinsky and Kowalik 2007).

In the meantime, the first Arctic coupled (Koenigk et al. 2011) and uncoupled (Rinke and Dethloff 2008) climate scenario projections were performed. Current regional Arctic modeling projects aim to integrate non-classic components such as vegetation and permafrost. By doing so, the models moved toward being Arctic system models rather than Arctic climate models (Roberts et al. 2011). Examples of such efforts are the Swedish ADSIMNOR project (ADSIMNOR 2014) and the US project RACM (Cassano et al. 2011), which are typical combinations of climate and system modeling with impact research.

The ACIA was carried out between 2000 and 2004, following a period of preparation and discussion of means and ends in the 1990s. Initiated by "boundary organizations" (Gieryn 1982) such as AMAP and IASC, the formal decision was made by the Arctic Council. A central question is what role this exercise at the science-policy interface played in the understanding of Arctic climate change when compared with increasing scientific activities to improve the models of global climate change.

At one level, it played a very limited role. Early in the ACIA process, the use of regional climate models for the assessment of impacts was discussed, but scientists in the assessment argued that these were not sufficiently developed and instead global models were combined with various efforts of statistical downscaling. Nevertheless, the report featured one whole chapter on modeling. A major message was the need for better representation of the Arctic Ocean and that 'the amplification of global-model systematic errors in regional models present[ed] a serious challenge to future regional model developments' (Kattsov and Källén 2005: 140).

Prominent in the ACIA chapter on modeling was also a discussion about climate variability and a conclusion that climate variability in the Arctic was much more important than previously thought. This conclusion was also at the center of a major discussion during the assessment process about whether one could statistically attribute Arctic climate change to anthropogenic drivers, which was ultimately decided in the negative by the lead authors of the chapters on modeling and the chapter on observations of past and current climate (Nilsson 2007).

Models thus became a tool for analyzing one of the key questions in climate policy discussions at the time—was Arctic climate change anthropogenic or natural? A question that had caused considerable controversy in the writing of the IPCC 2001 summary for policy makers (Nilsson and Döscher 2013). The discussion on variability was mainly limited to the scientific process and thus less visible when the ACIA results were communicated to a wider audience, where the message of ongoing and imminent change came to dominate.

The ACIA built on the IPCC and was also intended to feed into future IPCC assessments, but this was an assessment of published literature, not a research project, and it could not deliver further knowledge about the processes that created uncertainties regarding future climate change in the Arctic, such as the dynamics of sea ice. However, it served to change the framing of Arctic climate change from one where its importance was related mainly to uncertainties in global climate models to one in which the Arctic voice was heard in its own right. This shift becomes evident when the ACIA is compared with earlier IPCC assessments, where the voices of Arctic residents, especially indigenous peoples, became heard in ways that they had not before. The difference from the IPCC 2001 assessment is striking; it can be argued that the ACIA report, through its framing of the issue, brought a human face to Arctic climate change (Nilsson 2007).

This shift is significant in relation to how the Arctic region was seen in the public and scientific discourses. Half a decade into the new century and the Arctic voices were highlighted and their observations included as part of the understanding of climate change, in contrast to the previous practice of the Arctic being used as a laboratory for scientists from outside the region. This shift was closely linked with the political context brought by the Arctic Council, and the fact that indigenous peoples had a seat at the table together with nation states and a platform for taking part in scientific assessments. Organizations of indigenous peoples, the Inuit organization ICC in particular, played a significant role in formulating the scope of the ACIA and also in publicizing its results (Nilsson 2009b; Watt-Cloutier et al. 2006).

Within the Arctic Council, controversies surrounding the ACIA led to a lengthy stalemate regarding follow-up activities. One of the few timely follow-up events was a workshop on the application of climate scenarios in the local Arctic climate, which focused almost exclusively on statistical downscaling (AMAP 2007). Eventually the need for a follow-up to the ACIA became more important than trying to avoid the political sensitivities connected to climate policy. As the chairmanship moved to Norway in late

2006, attention to climate change was included as a priority in the joint program for the Norwegian, Danish, and Swedish chairmanships and in April 2008, the Arctic Council approved of a new project called "Snow, Water, Ice and Permafrost in the Arctic" (SWIPA). The goal was to assess current scientific information on changes in the Arctic cryosphere, including the impacts exerted by climate change. One of the messages in the assessment report was that the decline of sea ice plays a major role in the amplification of warming in the Arctic (AMAP 2011).

More recently, adaptation to climate change has become an important part of the political and scientific agenda of the Arctic Council, including the assessment Adaptation Action for a Changing Arctic (AACA). The major role of modeling has shifted toward providing information about potential impacts of climate change within the Arctic. However, it is also increasingly clear that impacts need to be assessed in the larger context of social and political complexities of adaptation (Hovelsrud and Smit 2010), and modeling efforts are thus not as exclusively important as they had been portrayed in earlier efforts to understand impacts. This more complex understanding also comes out in the polar chapter of the most recent IPCC report (Larsen et al. 2014).

While the political controversies within the Arctic Council led to a loss of momentum for its work on climate change after the ACIA, the scientific community was in the process of large-scale mobilization for the International Polar Year 2007–2008, an endeavor with long historical standing and traditionally of political importance due precisely to its international character. The plans included several studies of Arctic sea ice, such as the large EU-funded DAMOCLES (Developing Arctic Modeling and Observing Capabilities for Long-term Environmental Studies) as well as new efforts from a number of space agencies to launching satellites in order to acquire new data. In this effort, too, voices of Arctic indigenous peoples became prominent (Krupnik et al. 2011), further strengthening the norms about knowledge production in the Arctic, away from the trope of the Arctic as a laboratory or field site for outside actors toward a focus on the coproduction of knowledge by scientists and people living in the region.

The increased understanding of the rapid warming that has taken place in the Arctic in recent years is partly connected to a reanalysis of global climate models that have been optimized with the help of observational data (Edwards 2010). Even if additional mechanisms may contribute to Arctic amplification, such as changes in cloudiness, atmospheric humidity, and large-scale atmospheric and ocean circulation, "best guess" combinations of observations completed by constrained model simulations point to changes in sea ice cover as the primary mechanism behind Arctic amplification (e.g. Serreze et al. 2009; Screen and Simmonds 2010). Moreover, analyses of the observed changes in the Arctic quite clearly show a present-day warming trend. Despite the characteristically large internal variability in the Arctic, which earlier had prevented the ACIA authors from drawing conclusions about the attribution of Arctic climate change, it has now become possible to attribute the warming trend to anthropogenic climate forcing (Gillett et al. 2008; Nilsson and Döscher 2013). Various studies also emphasize that Arctic amplification can be observed during the most recent years (Pithan and Mauritsen 2014).

Conclusion

The emergence of new cultures of prediction in the Arctic, expressed in new institutional forms, computational techniques, and political relationships has had significant effects on broader scientific and cultural discourses of climate change. Climate modeling has played an important role in creating a picture of the Arctic as part of the world rather than an exceptional space, and has thus come to challenge the old trope of exceptionalism. In parallel, we have seen Arctic voices claiming space in the discussion about Arctic climate change in ways that challenge tropes of the Arctic as a field site for science and laboratory for outside interests. This latter trend is connected to contemporary political developments, in which Arctic cooperation—scientific and political—has created an opportunity for people in the region to formulate their needs for more knowledge about climate change, including for adaptation. Moreover, amidst the development of regional modeling, the idea of the Arctic as merely a bellwether of global warming has been complemented by a scientific formulation of polar amplification, which has rendered the area a new and more powerful position vis-à-vis the globe. It is noteworthy that this remote and sparsely populated region with its historically scarce data sets and late entry into the GCMs has come to be central in the global climate discussions and one on which scientific effort is focused. This can best be understood in relation to both its physical characteristics that are important for the climate and the political context of Arctic and climate science.

This coproduction between science and politics can be illustrated with the advent of the establishment of Arctic political cooperation. However, the political changes from Cold War conflict to cooperation have also contributed to a recognition of the need to understand Arctic climate change for the sake of people living in the region, rather than as a contribution to global political processes or as exclusive data delivered to global meteorological science.

In addition, the agency of sea ice itself entered the scene and began to play an active role in both political and scientific development. In the late summer of 2007, it was apparent that the extent of summer sea ice had reached a new record low, way below what most models had predicted. The sea ice minimum was important not only for scientists, who had to readjust their International Polar Year (IPY) project plans to take the new developments, but also politically, as it started a broader debate about the potential consequences of rapid Arctic climate change (Christensen et al. 2013).

The case of regional modeling in the Arctic illuminates issues of politics and agency that might be easier to deconstruct than in the corresponding global cases. The process of asserting anthropogenic climate change in the Arctic, partly through the changes in sea ice quality and extent, can be tied to national interests as articulated in the Arctic Council and subsequent scientific efforts. The coproduction mentioned above is locally grounded. The question of why the Arctic arrived to climate modeling so late is not simple to answer. It is clearly not a straight forward issue of the linear growth of scientific knowledge and improved modeling. The interplay between the enhanced capacity of science and the political, economic, and discursive framing conditions is crucial to understand the role of the Arctic in climate models. The geopolitical importance of the region and the organizational structuring of interests have enabled specific scientific research projects that have served both political and scientific ends. The resulting changed understanding of the Arctic ice in a global context has also challenged fundamental ideas on human agency and responsibility, and they continue to do so.

Note

1 The Arctic Council succeeded the Arctic Environmental Protection Strategy in 1996.

References

ACIA Arctic Climate Impact Assessment (2005) *Impacts of a Warming Arctic: Arctic Climate Impact Assessment*, Cambridge, MA: Cambridge University Press.

ADSIMNOR Advanced Simulation of Arctic Climate Change and Impact on Northern Regions (2014) *Homepage*, available from http://www.smhi.se/adsimnor, accessed 10 April 2015.

AEPS Arctic Environmental Protection Strategy (1993) *Arctic Environment*, (September 1993) Nuuk, Greenland: Second Ministerial Conference.

Agrawala, S. (1998a) "Context and Early Origin of the Intergovernmental Panel on Climate Change," *Climate Change* 39, pp. 605–620.

Agrawala, S. (1998b) "Structural and Process History of the Intergovernmental Panel on Climate Change," *Climate Change* 39, pp. 621–642.

Ahlmann, H. (1938) *Land of Ice and Fire: A Journey to the Great Iceland Glacier*, London: Routledge & Kegan Paul (Swedish original published in 1936).

Ahlmann, H. (1943) "Is och hav i Arktis," *Kungl. Svenska Vetenskapsakademiens Årsbok*, Stockholm: Almqvist and Wiksell, pp. 327–336.

AMAP Arctic Monitoring and Assessment Programme (2007) *Final Report of the Workshop on Adaptation of Climate Scenarios to Arctic Climate Impact Assessments*, (AMAP Report 14–16 May 2007, 4), Oslo, Norway: AMAP.

AMAP Arctic Monitoring and Assessment Programme (2011) *Snow, Water, Ice and Permafrost in the Arctic (SWIPA): Climate Change and the Cryosphere*, Oslo, Norway: AMAP.

Archer, C. and D. Scrivener (2000) "International Co-Operation in the Arctic Environment," M. Nuttall and T. V. Callaghan (eds.) *The Arctic: Environment, People, Policy*, Amsterdam, Netherlands: Harwood Academic Publishers, pp. 601–619.

ARCUS Arctic Research Consortium of the United States (2014) *Homepage*, available from http://www.arcus.org/search/sciencecoordination/development.php, accessed 10 April 2015.

Åtland, K. (2008) "Mikhail Gorbachev, the Murmansk Initiative, and the Desecuritization of Interstate Relations in the Arctic," *Cooperation and Conflict* 43(3), pp. 289–311.

Bitz, C. M., K. M, Shell, P. R. Gent, D. A. Bailey, G. Danabasoglu, K. C. Armour, M. M. Holland and J. T. Kiehl (2012) "Climate Sensitivity of the Community Climate System Model, Version 4," *Journal of Climate* 25, pp. 3053–3070.

Bolin, B., B. R. Döös, and J. Jäger (eds. 1986) *The Greenhouse Effect: Climate Change and Ecosystems*, Chichester, UK: John Wiley and Sons.

Bravo, M. T. (2009) "Preface: Legacies of Polar Science," J. Shadian and M. Tennberg (eds) *Final Frontiers or Global Laboratory? The Interface Between Science and Politics in the International Polar Year (IPY)*, Aldershot, UK: Ashgate.

Broecker, W. (1975) "Climatic Change: Are We on the Brink of a Pronounced Global Warming?" *Science New Series* 189(4201), pp. 460–463.

Broecker, W. (1987) "Unpleasant Surprises in the Greenhouse?" *Nature* 328, pp. 123–136.

Bryan, K. and M. D. Cox (1967) "A Numerical Investigation of the Oceanic General Circulation," *Tellus* 19(1), pp. 54–80.

Bryan, K., S. Manabe and R. C. Pacanowski (1975) "A Global Ocean-Atmosphere Climate Model. Part II. The Oceanic Circulation," *Journal of Physical Oceanography* 5(1), pp. 30–46.

Budyko, M. I. (1969) "The Effect of Solar Radiation Variations on the Climate of the Earth," *Tellus* 21, pp. 611–619.

Carey, M. (2010) *In the Shadow of Melting Glaciers: Climate Change and Andean Society*, Oxford, UK: Oxford University Press.

Cassano, J. J., M. E. Higgins and M. W. Seefeldt (2011) "Performance of the Weather Research and Forecasting (WRF) Model for Month-long Pan-Arctic Simulations," *Monthly Weather Review* 139(11), pp. 3469–3488.

Christensen, M., A. E. Nilsson and N. Wormbs (eds., 2013) *Media and the Politics of Arctic Climate Change: When the Ice Breaks*, New York, NY: Palgrave MacMillan.

CMIP Coupled Model Intercomparison Project (2014) *Homepage*, available from http://cmip-pcmdi.llnl.gov, accessed 10 April 2015.

Corell, R. (2004) *Interview of Robert Corell by A. Nilsson*, 24 March 2004.

Crawford, E. (1992) *Nationalism and Internationalism in Science, 1880–1939: Four Studies of the Nobel Population*, Cambridge, MA: Cambridge University Press.

Crawford, E. (1998) "Arrhenius' 1896 Model of the Greenhouse Effects in Context," H. Rodhe and R. Charlson (eds) *The Legacy of Svante Arrhenius: Understanding of the Greenhouse Effect*, Stockholm, Sweden: Royal Academy of Sciences and Stockholm University, pp. 21–32.

Cruikshank, J. (2005) *Do Glaciers Listen? Local Knowledge, Colonial Encounters, and Social Imagination*, Vancouver, Canada: UBC Press.

Dansgaard, W., H. B. Clausen, N. Gundestrup, C. U. Hammer, S. F. Johnson, P. M. Kristinsdoff and N. Reeh (1982) "A New Greenland Deep Ice Core," *Science* 218, pp. 1273–1277.

Dickinson, R. E. (1986) "How Will Climate Change?" B. Bolin, B. R. Döös, J. Jäger et al. (eds.) *The Greenhouse Effect, Climatic Change, and Ecosystems*, New York, NY: John Wiley, pp. 206–270.

Doel, R. E. (2003) "Constituting the Postwar Earth Sciences The Military's Influence on the Environmental Sciences in the USA After 1945," *Social Studies of Science* 33(5), pp. 635–66.

Doel, R. E., R. M. Friedman, J. L. Lajus, S. Sörlin and U. Wråkberg (2014) "Strategic Arctic Science: National Interests in Building Natural Knowledge before, during, and after World War II," *Journal of Historical Geography* 42, pp. 60–80.

Edwards, P. N. (2004a) "'A Vast Machine': Standards as Social Technology," *Science* 304, pp. 827–828.

Edwards, P. N. (2004b) "Representing the Global Atmosphere," C. A. Miller and P. N. Edwards (eds.) *Changing the Atmosphere: Expert Knowledge and Environmental Governance*, Cambridge, MA: MIT Press, pp. 31–65.

Edwards, P. N. (2010) *A Vast Machine: Computer Models, Climate Data, and the Politics of Global Warming*, Cambridge, MA: MIT Press.

Elzinga, A. (1993) "Antarctica: The Construction of a Continent by and for Science," E. Crawford, T. Shinn and S. Sörlin (eds.) *Denationalizing Science: The Contexts of International Scientific Practice*, (Sociology of the Sciences Yearbook 16), Dordrecht, Boston, London: Kluwer, pp. 73–106.

Endfield, G. H. (2011) "Re-culturing and Particularising Climate Discourses: Weather, Identity and the Work of Gordon Manley," *Osiris* 26, pp. 142–162

Franz, W. E. (1997) *The Development of an International Agenda for Climate Change: Connecting Science to Policy*, (ENRP Discussion Paper E-97-07), Cambridge, MA: Kennedy School of Government, Harvard University.

Gates, W. L., A. Henderson-Sellers, G. J. Boer et al. (1995) "Climate models – evaluation," J. T. Houghton, L. G. Meiro Filho, B. A. Callander et al. (eds.) *Climate Change 1995: The Science of Climate Change, Contribution of Working Group I to the Second Assessment Report of the Intergovernmental Panel on Climate Change*, Cambridge, MA: Cambridge University Press, Chapter 5.

Genthon, C., J. M. Barnola, D. Raynaud, C. Lorius, J. Jouzel, N. I. Barkov, Y. S. Korotkevich and V. M. Kotlyakov (1987) "Vostok Ice Core: Climatic Response to CO_2 and Orbital Forcing Changes Over the Last Climatic Cycle," *Nature* 329, pp. 414–418.

Gieryn, T. F. (1983) "Boundary-work and the Demarcation of Science from Non-science: Strains and Interests in Professional Ideologies of Scientists," *American Sociological Review* 48(6), pp. 781–795.

Gillett, N. P., D. A. Stone, P. A. Stott, T. Nowaza, A. Yu. Karpechko, G. C. Hegerl, M. F. Wehner and P. D. Jones (2008) "Attribution of Polar Warming to Human Influence," *Nature Geoscience* 1, pp. 750–754.

Griffies, S. M., C. Böning, F. O. Bryan, E. P. Chassignet, R. Gerdes, H. Hasumi, A. Hirst, A. -M. Treguier and D. Webb. (2000) "Developments in ocean climate modelling," *Ocean Modelling* 2(3), pp. 123–192.

Harper, K. (2008) *Weather by the Numbers: The Genesis of Modern Meteorology*, Cambridge, MA: MIT Press.

Heininen, L. (2004) "Circumpolar International Relations and Geopolitics," *Arctic Human Development Report*, Akureyri: Stefansson Arctic Institute, pp. 207–227.

Heymann, M. (2009) "Klimakonstruktionen. Von der klassischen Klimatologie zur Klimaforschung," *NTM. Journal of the History of Science, Technology and Medicine* 17(2), pp. 171–197.

Heymann, M. (2010) "The Evolution of Climate Ideas and Knowledge," *Wiley Interdisciplinary Reviews Climate Change* 1(3), pp. 581–597.

Hibler, W. D. (1979) "A Dynamic Thermodynamic Sea Ice Model," *Journal of Physical Oceanography* 9, pp. 815–846.

Houghton, T., G. J. Jenkins and J. J. Ephraums (eds., 1990) *Scientific Assessment of Climate Change – Report of Working Group I*, Cambridge, MA: Cambridge University Press.

Hovelsrud, G. K. and B. Smit (eds., 2010) *Community Adaptation and Vulnerability in Arctic Regions*, Dordrecht, Netherlands: Springer.

Hulme, M. (2009) *Why We Disagree about Climate Change: Understanding Controversy, Inaction and Opportunity*, Cambridge, MA: Cambridge University Press.

Jasanoff, S. (ed., 2004) *States of Knowledge: The Co-Production of Science and Social Order*, London, New York: Routledge.

Jasanoff, S. and B. Wynne (1998) "Science and Decision Making," S. Rayner and E. L. Malone (eds.) *Human Choice & Climate. Vol. 1. The Societal Framework*, Columbus, OH: Battelle Press, pp. 1–87.

Johnson Theutenberg, B. (1982) "Polarområdena - Politik Och Folkrätt," Swedish Royal Academy of Sciences (ed.) *Polarforskning: Förr, Nu Och i Framtiden*, Stockholm, Sweden: Swedish Royal Academy of Sciences, pp. 40–55.

Kattsov, V. M. and E. Källén (2005) "Future Climate Change: Modeling and Scenarios for the Arctic," ACIA (ed.) *Arctic Climate Impact Assessment*, Cambridge, MA: Cambridge University Press, pp. 99–150.

Knipowitsch, N. M. (1921) "O termicheskikh usloviiakh Barentseva moria v kontse maia 1921 goda," *Bulleten' Rossiiskogo Gidrologicheskogo Instituta* 9, pp. 10–12.

Koenigk, T., R. Döscher and G. Nikulin (2011) "Arctic Future Scenario Experiments with a Coupled Regional Climate Model," *Tellus A* 63, pp. 69–86.

Krupnik, I., M. A. Lang, and S. E. Miller (ed., 2009) *Smithsonian at the Poles: Contributions to International Polar Year science*, Washington, DC: Smithsonian Institution Scholarly Press.

Krupnik, I., C. Aporta, S. Gearheard, S. Laidler, G. J. Kielsen and L. Holm (2010) *SIKU: Knowing Our Ice: Documenting Inuit Sea Ice Knowledge and Use*, Dordrecht, Netherlands: Springer.

Krupnik, I., I. Allison, R. Bell, P. Culter, D. Hik, J. López-Martínez, V. Rachold, E. Sarukhanian and C. Summerhayes (2011) *Understanding Earth's Polar Challenges: International Polar Year 2007–2008*, available from http://library.arcticportal.org/id/eprint/1211, accessed 14 Feb 2017.

Kuhn, T. S. (1962) *The Structure of Scientific Revolutions*, Chicago, IL: The University of Chicago Press.

Lajus J. and S. Sörlin (2014) "Melting the Glacial Curtain: The Politics of Scandinavian-Soviet Networks in the Geophysical Field Sciences Between Two Polar Years, 1932/33–1957/58," *Journal of Historical Geography* 42, pp. 44–59.

Larsen, J. N., O. A. Anisimov, et al. (2014) "Polar regions," C. B. Field, V. R. Barros, D. J. Dokken et al. (eds.) *Climate Change 2014: Impacts, Adaptation, and Vulnerability. Part B: Regional Aspects, Contribution of Working Group II to the Fifth Assessment Report of the Intergovernmental Panel on Climate Change*, Cambridge, MA: Cambridge University Press, Chapter 28.

Launius, R. D., J. R. Fleming and D. H. DeVorkin (eds., 2010) *Globalizing Polar Science: Reconsidering the International Polar and Geophysical Years*, New York, NY: Palgrave Macmillan.

Macdougall, D. (2004) *Frozen Earth: The Once and Future Story of Ice Ages*, Berkeley, CA: University of California Press.

Manabe, S. (1969) "Climate and the Ocean Circulation, 1. The Atmospheric Circulation and the Hydrology of the Earth's Surface," *Monthly Weather Review* 97, pp. 937–805.

Manabe, S. and K. Bryan (1969) "Climate Calculations with a Combined Ocean-Atmosphere Model," *Journal of Atmospheric Science* 26, pp. 786–789.

Manabe, S., K. Bryan, and M. J. Spelman (1975) "A Global Ocean-Atmosphere Climate Model: Part I. The Atmospheric Circulation," *Journal of Physical Oceanography* 5(1), pp. 3–29.

Manabe, S. and R. Wetherald (1975) "The Effects of Doubling the CO2 Concentration on the Climate of General Circulation Model," *Journal of Atmospheric Science* 32(1), pp. 3–15.

Manabe, S. and R. J. Stouffer (1980) "Sensitivity of a Global Climate Model to an Increase of CO2 Concentration in the Atmosphere," *Journal of Geophysical Research* 85(C10): 5529–5554.

McAveney, B. J. (2001) "Model evaluation," J. T. Houghton, Y. Ding, D. J. Griggs, N. Noguer, P. J. van der Linden, D. Xiaosu, K. Maskell and C. A. Johnson (eds.) *Climate Change 2001: The Scientific Basis*, Cambridge, MA: Cambridge University Press, Chapter 8.

Miller, C. A. (2004) "Climate Science and the Making of Global Political Order," S. Jasanoff (ed.) *States of Knowledge. The Co-Production of Science and Social Order*, London, New York, NY: Routledge, pp. 46–66.

Miller, C. A. and P. N. Edwards (eds., 2001) *Changing the Atmosphere: Expert Knowledge and Environmental Governance*, Cambridge, MA: MIT Press.

National Research Council (1986) *The National Climate Program: Early Achievements and Future Directions*, (Board on Atmospheric Sciences and Climate, Commission on Physical Sciences, Mathematics, and Resources), Washington, DC: National Research Council.

Nilsson, A. E. (2007) *A Changing Arctic Climate: Science and Policy in the Arctic Climate Impact Assessment*, (Dept. of Water and Environmental Studies), Linköping, Sweden: Linköping University Press.

Nilsson, A. E. (2009a) "A Changing Arctic Climate: More Than Just Weather," J. Shadian and M. Tennberg (eds.) *Final Frontiers or Global Laboratory? The Interface Between Science and Politics in the International Polar Year (IPY)*, Aldershot, UK: Ashgate, pp. 9–33.

Nilsson, A. E. (2009b) "Arctic Climate Change: North American Actors in Circumpolar Knowledge Production and Policymaking," H. Selin and S. D. VanDeveer (eds.) *Changing Climates in North American Politics: Institutions, Policymaking and Multilevel Governance*, Cambridge, MA: MIT Press, pp. 77–95.

Nilsson, A. E. and R. Döscher (2013) "Signals from a Noisy Region," M. Christensen, A. E. Nilsson and N. Wormbs (eds.) *Media and the Politics of Arctic Climate Change: When the Ice Breaks*, London: Palgrave Macmillan, pp. 93–113.

Paglia, E. (2016) "The Teleconnected Arctic: Ny-Ålesund, Svalbard as Scientific and Geopolitical Node" E. Paglia, *The Northward Course of Anthropocene: Transformation, Temporality and Telecoupling in a Time of Environmental Crisis*, (dissertation), Stockholm, Sweden: KTH Royal Institute of Technology.

Pithan, F. and T. Mauritsen (2014) "Arctic Amplification Dominated by Temperature Feedbacks in Contemporary Climate Models," *Nature Geoscience* 7, pp. 181–184.

Proshutinsky, A. and Z. Kowalik (2007) "Preface to Special Section on Arctic Ocean Model Intercomparison Project (AOMIP) Studies and Results," *Journal of Geophysical Research* 112, C04S01.

Rinke, A. and K. Dethloff (2008) "Simulated Circum-Arctic Climate Changes by the End of the 21st Century," *Global Planet Change* 62, pp. 173–186.

Roberts, A., J. Cherry, R. Döscher, S. Elliott, L. Sushama (2011) "Exploring the Potential for Arctic System Modeling," *Bulletin of American Meteorological Society* 92(2), pp. 203–206.

Ross, A. (1991) *Strange Weather: Culture, Science and Technology in the Age of Limits*, New York, NY: Verso.

Rothrock, D. A., Y. Yu and G. A. Maykut (1999) "Thinning of the Arctic sea-ice cover," *Geophysical Research Letters* 26, pp. 3469–3472.

Rummukainen, M. (2010) "State-of-the-art with Regional Climate Models," *Wiley Interdisciplinary Reviews Climate Change* 1, pp. 82–96.

Schlesinger, M. E. (1988) *Physically-Based Modelling and Simulation of Climate and Climatic Change: Part 1*, Amsterdam, Netherlands: Kluwer Academic Publishers.

Screen, J. A. and I. Simmonds (2010) "The Central Role of Diminishing Sea Ice in Recent Arctic Temperature Amplification," *Nature* 464, pp. 1334–1337.

Serreze, M. C. and J. A. Francis (2006) "The Arctic Amplification Debate," *Climatic Change* 76(3–4), pp. 241–264.

Serreze, M. C. and J. A. Francis (2006a) "The Arctic on the Fast Track of Change," *Weather* 61(3), pp. 65–69.

Serreze, M. C., A. Barrett, J. C. Stroeve, D. N. Kindig and M. M. Holland (2009) "The Emergence of Surface-based Arctic Amplification," *The Cryosphere* 3, pp. 11–19.

Shadian, J. and M. Tennberg (eds., 2009) *Legacies and Change in Polar Sciences: Historical, Legal and Political Reflections on the International Polar Year*, Farnham: Ashgate.

Sörlin, S. (2009) "Narratives and Counter Narratives of Climate Change: North Atlantic Glaciology and Meteorology, ca 1930–1955," *Journal of Historical Geography* 35(2), pp. 237–255.

Sörlin, S. (2011) "The Anxieties of a Science Diplomat: Field Co-production of Climate Knowledge and the Rise and Fall of Hans Ahlmann's 'Polar Warming'," J. R. Fleming and V. Jankovich (eds.) *Osiris 26: Revisiting Klima*, Chicago, IL: University of Chicago Press, pp. 66–88.

Sörlin, S. (2013) "Ice Diplomacy and Climate Change: Hans Ahlmann and the Quest for a Nordic Region Beyond Borders," S. Sörlin (ed.) *Science, Geoplitics and Culture in the Polar Region – Norden beyond Borders*, Farnham: Ashgate, pp. 23–54.

Sörlin, S. and J. Lajus (2013) "An Ice Free Arctic Sea? The Science of Sea Ice and Its Interests," M. Christensen, A. E. Nilsson and N. Wormbs (eds.) *Media and the Politics of Arctic Climate Change: When the Ice Breaks*, New York, NY: Palgrave MacMillan, pp. 70–92.

Sundberg, M. (2011) "The Dynamics of Coordinated Comparisons: How Simulationists in Astrophysics, Oceanography and Meteorology Create Standards for Results," *Social Studies of Science* 41(1), pp. 107–125.

Tennberg, M. (2000) *Arctic Environmental Cooperation: A Study in Governmentability*. Hants, Burlington, UK: Ashgate Publishing Company.

Watt-Cloutier, S., Fenge, T. and P. Crowley (2006) "Responding to Global Climate Change: The view of the Inuit Circumpolar Conference on the Arctic Climate Impact Assessment," *2° is too much! Evidence and Implications of Dangerous Climate Change in the Arctic*, Oslo, Norway: WWF, pp. 57–68.

Weart, S. R. (2008) *The Discovery of Global Warming*, Cambridge, MA: Harvard University Press.

Weatherhead, E. C. (1998) "Climate Change, Ozone, and Ultraviolet Radiation," *AMAP Assessment Report. Arctic Pollution Issues*, Oslo: AMAP, pp. 717–774.

Wright, J. K. (1953) "The Open Polar Sea," *Geographical Review* 43(3), pp. 338–365.

Young, O. R. (1998) *Creating Regimes: Arctic Accords and International Governance*, Ithaca, NY: Cornell University Press.

Zubov, N. N. (1933) "The Circumnavigation of Franz Josef Land," *Geographical Review* 23(3), pp. 394–401 and p. 528.

10 From predictive to instructive

Using models for geoengineering

Johann Feichter and Markus Quante

> *"Man masters nature not by force, but by understanding."*
> (J. Bronowski, Polish mathematician)

Introduction

The old dream of mankind to influence weather and climate has come true at least in the virtual world of computer simulations. When it became apparent that humans alter the climate system, it was just a short step further to launch proposals to alter the climate intentionally, or to manage or even optimize the Earth system. With the advent of fast computers and enormous increases in the capabilities of measurement devices, tools were developed to simulate the complex Earth system and to answer "What if?" questions. Schellnhuber (1999) called this increasing understanding of the Earth system and the development of concepts for global environmental management the "Second Copernican Revolution." Although geoengineering concepts are mostly in their infancy, field experiments have been conducted and the deployment of geoengineering is hotly contested.

Analyses of complex nonlinear systems such as the climate system require mathematical models as well as powerful computers. During the past decades, numerical climate models have been developed and applied to gain an understanding of the climate system's behavior and to assess the potential effects of anthropogenic activities on the climate. Within the framework of the Intergovernmental Panel on Climate Change (IPCC) assessments, climate models have also simulated a range of possible future scenarios describing different economic pathways. As a consequence, these scenario simulations were followed by efforts to use climate models to explore pathways to optimize the climate. Such ideas have been discussed for decades within the scientific community, but only recently have they entered the political arena.

First, we present a brief outline of the history of geoengineering ideas and describe the main methods under discussion. Next, we discuss the limitations of experiments designed to serve in advising policy makers and the potential of numerical climate models to explore the potential of various geoengineering proposals. If models are used for policy advice, which demands must be met in terms of accuracy and predictive skills?

The history of geoengineering

The idea of geoengineering first reached a wider public in the 1940s and 1950s when there was a strong belief that modern techniques would allow us to control all aspects of life, although the issue had already been under discussion earlier within the scientific community (Keith 2000). The idea to intentionally interfere with the atmosphere or components of the climate system is not a new one. As early as 1908, Svante Arrhenius proposed—in strict contrast to today's objectives—deliberately enhancing the greenhouse effect by burning more fossil fuels to enhance agricultural productivity (Arrhenius 1908). Over many centuries, scholars and mythologers proposed schemes for weather modification, most of them dealing with rainmaking. A first "scientifically" based rainmaking theory emerged around 1830 (Fleming 2006). James P. Espy proposed to enhance precipitation by lighting huge fires, thus stimulating thermally induced convective updrafts. A more scientifically grounded method of weather modification began in 1940, with discoveries in the field of cloud seeding by a group of scientists around Irving Langmuir at the General Electric Corporation in the United States (Fleming 2006). They discovered that seeding supercooled clouds with chemicals having a similar crystallographic structure as ice initiates precipitation. The intention of cloud seeding was to enhance precipitation in dry areas, or to precipitate out clouds before strong storms or hail could develop (e.g. Project Stormfury, a joint venture of the United States Department of Commerce and the United States Navy, which ran from 1962 to 1983, and attempted to weaken tropical cyclones by cloud seeding). But rainmaking and suppressing storms and hail was not the only focus of weather modification. Soon military agencies in the United States devoted significant funds to research on what came to be called "climatological warfare" (Keith 2000; Fleming 2007). Details of this research remained concealed behind a veil of secrecy. Similar research activities took place in the Soviet Union and later in China, but information about these military programs is even more fragmentary than that about the US programs.

The famous mathematician, John von Neumann, who organized meetings with leading US scientists to explore the possibility of weather modification, in a 1955 *Fortune* magazine article at the height of the Cold War, foresaw forms of climatic warfare yet unimagined (Weart 2011; see also Rosol in this volume). In fact, in the 1960s the US government had secretly been spending a large amount of money on experiments in climatological warfare. The most famous of these experiments was conducted under the acronym POPEYE between 1967 and 1972 by the Department of Defense, which directed extensive cloud seeding over the Ho Chi Minh Trail in Vietnam, with the intention of flooding the North Vietnamese Army's supply line (Weart 2011). This program is relevant to the topic of geoengineering because, two years after the program's end, the secret was exposed by journalists and the public learned about the operation. The succeeding international discussion led to the United Nations (UN) Convention on the Prohibition of Military or Any Other Hostile Use of Environmental Modification Techniques (ENMOD) (drafted Dec. 1976, effective Oct. 1978). ENMOD and the UN Convention on Biological Diversity (CBD, effective Dec. 1993, not ratified by the US) are the only existing international agreements that partly cover geoengineering deployment. In article I, ENMOD points out that "each State Party to this Convention undertakes not to engage in military or any other hostile use of environmental modification techniques" (ENMOD 1978). The tenth meeting of the Conference of the Parties to the Convention on Biological Diversity in October 2010 adopted decisions on climate-related geoengineering and its impacts on achieving the objectives of the CBD. However, the resolution is restricted to banning measures that affect biodiversity, like ocean fertilization. The substance of the ENMOD Convention was reaffirmed in the United Nations Framework Convention on Climate Change (UNFCCC 1992) signed at the 1992 Earth Summit in Rio de Janeiro, which adopted principle 21 from the Declaration of the United Nations Conference on the Human Environment (UNCHE):

> States have [...] in accordance with the Charter of the United Nations and the principles of international law, the [...] responsibility to ensure that activities within their jurisdiction or control do not cause damage to the environment of other States or of areas beyond the limits of national jurisdiction.
>
> (UNCHE 1972)

In the next decades, funding levels for weather modification increased, but they then experienced a strong decline in the early 1980s, one of the reasons being that the potential of the method was oversold to funding agencies and to the public (see section "The crux of experiments").

Alongside the warfare proposals several proposals for the modification of weather and climate on a large scale were put forward, which sought to convert a greater percentage of incoming solar radiation into "useful energy," with the overall goal of warming certain regions of the globe. The ideas of the time are summarized by Wexler (1958). They mainly considered blackening deserts and polar ice caps or the formation of sufficiently thick ice clouds to reduce outgoing infrared radiation. Presumably, studies were conducted on climatological warfare in the Cold War Soviet Union as well, but little is known (Weart 2011). Some non-military ideas to shape the landscape and improve harsh climate conditions based on hydraulic engineering became public. A booklet published by Rusin and Flit (1960), with the telling title *Man Versus Climate*, described several huge programs for "improving" the climate. Most notably, proposals are depicted for diverting the flow of major northern rivers to the Russian wheat fields, or from the Mediterranean to irrigate Central Asian deserts, and geo-hydrographical schemes like damming the Bering Strait to warm the Arctic. Especially in reaction to massive Soviet plans, the famous British climatologist, H.H. Lamb, published a critical review paper titled "Climate-engineering schemes to meet a climatic emergency" (Lamb, 1971; see also Martin-Nielsen in this volume). Lamb highlighted for the first time the possible unintended and undesirable side effects of the proposed schemes on the distribution of climates over many parts of the Northern Hemisphere. He also introduced the term "climate engineering," which today is preferred by many scientists over "geoengineering," an expression coined by Marchetti (1977).

A paradigm shift occurred around the mid-1960s. While earlier climate modification schemes aimed at changing conditions towards a desired new, "improved," mainly warmer climate, the later proposals had the goal of counteracting anthropogenic climate change and thereby preserving the climate in its actual state. In 1965, President Johnson received the first ever presidential briefing on the dangers of carbon dioxide- (CO_2-) induced climate change, and the sole suggested response to the impact of expected global warming was climate engineering. The possibility of reducing fossil fuel use is not mentioned in the document (Weart 2011). The only scheme analyzed in the report was the modification of the albedo of the Earth by dispersing buoyant reflective particles over large areas of the tropical oceans with the aim of changing the planetary albedo by 1 percent.

Aside from the presidential report, several additional ideas to counteract expected global warming by climate system intervention were published during the period, often in the so-called "gray literature." Kellogg and Schneider (1974) discuss several of these potential climate engineering schemes and for the first time extensively point to possible hazards in connection with climate modification, calling it dangerous to pursue any large-scale climate control scheme until long-term effects could be predicted with acceptable assurance. The most famous early proposal formulated in the 1970s, which is still the basis of some of today's solar radiation management (SRM) schemes, was proposed by Mikhail Budyko, who suggested increasing the global albedo to counter carbon dioxide-induced warming by mimicking the action of intense volcanic eruptions, which inject large amounts of sulfur dioxide into the stratosphere (Budyko 1974, 1977). Over the following years every on occasion a publication addressing climate engineering, as a means of counteracting greenhouse gas-induced climate change, appeared in the literature. A few climate assessments were also compiled, some of them containing sections on deliberate climate modification–they are listed and discussed in Keith (2000). Among these assessments the most cited became the National Academy of Sciences report of 1992 (NAS 1992), which included a chapter on "geoengineering" that provided a detailed analysis of four different options: reforestation, ocean fertilization, albedo modification, and the removal of chlorofluorocarbons from the atmosphere. Concerning albedo modification schemes based on particle scattering in the upper atmosphere, it should be mentioned that the famous weapon system designer Edward Teller collaborated with his student Lowell Wood at Lawrence Livermore National Laboratory to re-examine this option (Teller et al. 1997). Instead of injecting sulfur dioxide, Teller and colleagues suggested the use of engineered alumina particles of nanometer size, which would also lower the impact on ozone chemistry in the stratosphere, a concern that had been formulated in some assessment studies.

In the following years, the controversial topic of climate engineering research receded into the background. Perhaps influenced by the strengthening of the ecological movement in the 1970s and 1980, there seemed to be a kind of self-imposed moratorium by scientists. That remained the case until Paul Crutzen, a Noble laureate for chemistry, started a lively debate with his paper published in 2006. He proposed to combat climate warming by injecting sulfur into the stratosphere. His proposal was motivated by the observed cooling induced by the Pinatubo eruption, as well as by the fact that international agreement on CO_2 emission cuts is difficult to achieve. Similar ideas were also developed earlier by a marine biogeochemist who wrote an article with the title "Insurance against a Bad Climate Trip". However, he did not publish it, after colleagues warned him not to open Pandora's box (pers. comm.). How skeptical the scientific community was about touching on this taboo topic publicly is further demonstrated by the fact that the editor of the journal *Climatic Change*—the journal where Paul Crutzen submitted his paper—asked five other scientists in 2006 to comment on Crutzen's proposal and published the comments in the same issue. From the very titles of their comments we find a broad spectrum of opinions about the geoengineering issue: "Encouraging research and overseeing implementation" by the editor R. J. Cicerone (2006), "Geoengineering climate change: Treating the symptom over the cause" by J. T. Kiehl (2006), "Geoengineering to confine climate change: Is it at all feasible?" by L. Bengtsson (2006, the tone of his article leans more toward "no"), "Geoengineering: Worthy of cautious evaluation?" by M. C. MacCracken (2006, in favor of regional geoengineering in the Arctic), and "The geoengineering dilemma: To speak or not to speak" by M. G. Lawrence (2006). Besides a vague feeling of unease and the attitude of environmentalists to be skeptical about the human desire to manipulate and control nature (Minteer 2012), the idea of geoengineering was not least contaminated by the many proposals to apply weather modification and geoforming for military purposes.

Three months after Crutzen's article was published, Ken Caldeira of Stanford University organized a workshop, sponsored by the NASA Ames Research Center and the Carnegie Institution of Washington, on "Management of Solar Radiation". Almost all participants in this workshop came from US institutions. NASA expected recommendations for research and development of geoengineering strategies. However, the workshop did not give any recommendations as to whether a method should be deployed. Instead, it was limited to defining the important scientific questions that could lessen uncertainty (Lane et al. 2006).

Proposed methods

The term "geoengineering" encompasses a wide range of distinct technology proposals. The proposed schemes can be subdivided into two categories. On the one hand, there are those which aim to reduce the amount of solar radiation absorbed by the Earth via sun shading or artificially increasing the albedo of the Earth, and on the other hand there are those schemes which directly influence the outgoing long-wave radiation by extracting CO_2 from the atmosphere, thereby reducing the concentration of the most important greenhouse gas. The proposals addressing the shortwave part of the radiation spectrum are usually summarized under "solar radiation management" (SRM) and those aiming to reduce the CO_2 concentration in the atmosphere are collectively pooled into the category "carbon dioxide removal" (CDR). Since the greenhouse effect, the main driver of current global warming, is an effect acting on the long-wave side of the radiation spectrum, CDR is the approach that tackles the root of the problem. Implied distorted compensation patterns induced by SRM methods to counter a long-wave forcing may lead, directly or via feedbacks, to complex changes in regional circulations, which again might have profound effects on the hydrological cycle (e.g. Hegerl and Solomon 2009).

To provide a brief overview, the currently most discussed geoengineering schemes from the two categories are listed in Table 10.1 along with brief explanations.

Underlying the geoengineering schemes in Table 10.1 there is a huge body of literature and technical proposals, too much to be introduced here. Instead, we refer readers to existing scientific reviews—some of them introducing a critical note—by the Royal Society (2009), Feichter and Leisner (2009), Vaughan and Lenton (2011), and Bellamy et al. (2012). A thorough discussion of the characterization of the different climate engineering methods in the context of mitigation of, and adaptation to, climate change can be followed in Boucher et al. (2014).

The proposed climate engineering schemes feature different efficiencies and time scales for research, implementation, effectiveness, and shutdown. In general, the SRM measures have shorter time scales than those of CDR. Possible side effects are speculated about, but they are not really known at present. Under discussion is, among other impacts, a potential reduction of rainfall in some regions of the world, or additional ozone depletion in the case of an increased stratospheric particle load. The worrying global change aspect of ocean acidification remains untouched if only SRM methods are considered without an accompanying reduction in carbon dioxide emissions. Overall, the science underlying many of the schemes is still in its infancy and possible side effects remain under-researched. A first comparison addressing the effectiveness of a set of climate engineering options using the common measure of radiative forcing potentials has been conducted by Lenton and Vaughan (2009). A preliminary risk-assessment study including a sort of ranking has been published by Boyd (2008). The comprehensive report of The Royal Society (Royal Society 2009) includes the attempt of an overall evaluation of proposed climate engineering schemes addressing effectiveness, affordability, timeliness, and safety. According to the authors of this report such an evaluation inevitably remains somewhat subjective due to the present incomplete state of knowledge. A less well-studied area of the potential side effects of climate engineering efforts is

Table 10.1 Geoengineering methods

Solar Radiation Management (SRM)	
Sunshades in space	Launch of very large mirrors or a giant number of smaller sunshades, which will be placed on the inner Lagrange point and redirect a fraction of the incoming sunlight before it enters the Earth system. A fleet of orbiting mirrors in the Earth's upper atmosphere also has been suggested.
Stratospheric aerosol	Injection of sulfur aerosols or designed particles into the upper stratosphere, using airplanes, balloons, long pipes, or projectiles. The aerosols alter the Earth's albedo by scattering a proportion of the incoming sunlight back into space, mimicking the effect of a strong volcanic eruption.
Cloud whitening	Spraying of small seawater droplets from many wind-driven vessels into the turbulent boundary layer underlying marine clouds. A small fraction of the droplets is thought to serve as additional condensation nuclei and increase the number of droplets, and thus the albedo of existing clouds.
Surface albedo enhancement	Artificially increasing the reflectivity of different natural or built surfaces (desert, grassland, cropland, human settlements).

Carbon Dioxide Removal (CDR)	
Land carbon sink enhancement	*Afforestation*: Increasing the amount of carbon extracted from the atmosphere and stored in biomass by converting non-forested into forested land.
	Bio-char: Production of fine charcoal by pyrolysis of biomass (bio-char) and subsequent long-term physical storage by burial (e.g. dug or plowed into soil to enhance nutrient content).
	Atmospheric carbon capture: Direct capture of CO_2 in air masses by using some form of scrubbing device with a chemical absorbent. The CO_2 is bound only lightly so that it can subsequently be released and transformed chemically before final storage.
Ocean carbon sink enhancement	*Ocean fertilization*: Systematic fertilization of ocean waters that lack certain nutrients to enhance phytoplankton productivity and consequently increase the sequestration of atmospheric CO_2 into deep water. Fertilization with iron, nitrogen or phosphorus has been suggested.
	Enhance upwelling: Mechanically lifting nutrient-rich water by free floating devices or tethered pipes to the less nutrient-rich surface to create additional carbon sinks.
	Increase ocean alkalinity: Addition of certain carbonate minerals to increase the alkalinity of the oceans and thereby enhance the ocean carbon sink; the lowered pH value exploits the carbonate chemistry in the ocean, allowing more carbon dioxide to be absorbed.

that of non-climatic impacts on ecosystems and biodiversity. Here, large scale CDR carries potentially severe consequences, which call for internationally coordinated studies (Williamson 2016).

The economic valuation of the different schemes is still an evolving field of research, but the range of related necessary financial means appears to be quite broad. While aerosol-based schemes in SRM were appraised as being "incredibly cheap" (Barret 2008)—although subsequent economic studies relativized the costs and came up with higher numbers (e.g. Rickels et al. 2011)—technical atmospheric capturing of CO_2 is estimated to be rather costly (House et al. 2011). The problem of evaluating the methods is extremely complex and scientists have often underestimated the impact of natural variability. Geoengineering research will inevitably face problems such as high natural variability, complex physics and feedbacks exerting perilous side effects. In addition, public attitudes about intentional changes to the environment have shifted and been modified by a new environmental ethic which often evokes skepticism about proposed geoengineering schemes.

The crux of experiments

Besides thought experiments, real experiments in the Arctic were also discussed during the NASA workshop. In particular, it was proposed to inject large amounts of aerosols into the Arctic lower stratosphere in summer. There the aerosols might persist for about six months and thus not impact winter ozone levels (enhanced ozone destruction is a possible side effect of injecting sulfate into the stratosphere). The aerosol-induced solar radiation reduction could promote an earlier icing-up of the Arctic Ocean, which would, in turn, induce further cooling because of the higher albedo of sea ice (MacCracken 2006). NASA was in favor of performing such reversible regional scale tests, but there was clearly some dissent between the organizers and some workshop participants. For instance, Alan Robock wrote an email criticizing an early draft of the workshop report:

> Although I think most people advocated conducting theoretical studies with computer models, I think most people were against actually doing experiments in the atmosphere.
>
> (email, Robock 2007)

In summary, there was broad agreement not to do experiments in the real world and not to develop geoengineering strategies, but instead to explore the potential of SRM and to study the possible perilous side effects through the means of numerical models.

However, it was not only scientists who did not feel at ease with performing real world geoengineering experiments. Sectors of the public also viewed geoengineering with skepticism. For instance, UK scientists obtained funding for a project (from 2010 to 2014) called Stratospheric Particle Injection for Climate Engineering (SPICE) that was to answer three questions (SPICE 2013).

How much, of what, needs to be injected where into the atmosphere to effectively and safely manage the climate system?
How do we deliver it there?
What are the likely impacts?

The project proposed both numerical simulations and real world experiments. Experiments planned for September 2011 to evaluate the simulations and test the technical equipment were first delayed by half a year, and then canceled in May 2012 because of concerns about the lack of government regulation regarding such geoengineering projects (thereluctantgeoengineer blog 2012).

Less public debate and attention arguably enabled ocean fertilization experiments at remote sites. There several experiments have been performed. However, such small-scale experiments—although the spatial scale increased from experiment to experiment—produced inconsistent results due to the great natural variability of the ocean-biosphere system. Therefore, future field experiments would need to cover larger spatial and temporal scales, enhancing the risk of adverse impacts and challenging the boundaries between experiment and geoengineering applications (Toussaint 2013). The problem of drawing conclusions from temporally and spatially limited experiments was put in a nutshell by James P. Espy (see section "The history of geoengineering") who

> in the summer of 1849 burned down twelve acres of timber in Fairfax County, Virginia [...] in the hope of producing an intense column of heated air, clouds, and artificial rain. [When Espy tried the experiment he] made sure that if rain was not produced, the failure could be attributed to unfavorable ambient conditions and not to any deficiency in his theory. The experiment ended in failure.
>
> (Fleming 2006: 6)

Generally, experiments in the real world have deficiencies in at least one of the key requirements: spatial and temporal coverage, accuracy of the accompanying measurements and number of parameters which are measured, and/or high costs. Moreover, not all experiments desirable from a scientific point of view can and should be performed. The use of numerical models offers a way out of these problems but at the price of uncertainty.

Geoengineering in a virtual world

A natural climate experiment took place on 15 June 1991, when the Philippine volcano Mount Pinatubo erupted, sending large amounts of sulfur dioxide and dust into the stratosphere up to an altitude of more than 25 km. This stratospheric aerosol shield reduced solar heating on Earth and resulted in a cooling of 0.5°C in the Northern hemisphere the year after the eruption. Hansen et al. (1992) were the first to use the Mount Pinatubo eruption to evaluate their climate model. They performed a series of model simulations immediately after the eruption and in January 1992 published a prediction of surface temperature changes close to the observed ones. In the following years, numerous model studies investigating the climatic impact of the Mount Pinatubo eruption were published. Thus, at the time Paul Crutzen published his paper, many climate models had the capability to qualitatively simulate the climatic consequences of stratospheric sulfur injections (Rasch et al. 2008) and, due to numerous studies about aerosol impacts on climate, of the seeding of marine clouds as well (Jones et al. 2009; Jones and Haywood 2012).

Within the framework of the Geoengineering Model Intercomparison Project (GeoMIP), twelve groups of modelers performed a set of standard-ized simulations using up-to-date climate and chemistry models to evaluate the effects of SRM (Kravitz et al. 2011). A special issue of the *Journal of Geophysical Research: Atmospheres* was published, containing fifteen papers on GeoMIP. The simulations show a compensation for the warming induced by green-house gases, but leave the tropics cooler and the poles warmer, and the tropical regions with less precipitation (Kravitz et al. 2013). All monsoon areas show reduced precipitation, manifested as reduced numbers of intense precipitation events (Tilmes et al. 2013). Analyses of all simulations indicate that differences in the microphysical representations of aerosols and in the parameterization of processes involved in carbon cycle feedbacks are the primary contributors to the differences among model simulations (Kravitz et al. 2014a). Thus, uncertainties in aerosol formation, growth, and deposition remain prominent limitations in geoengineering research. The lack of ability to accurately simulate some of the dynamic effects of volcanic eruptions further shadows the reliability of present day models when actual risk assessments are needed, like those that would be needed to advise policy before the actual deployment of methods, even in the case of smaller scale real world experiments. The GeoMIP community is quite active; inspired by the shortcomings of the available model experiments, a new experiment has been outlined (Tilmes et al. 2014). Significantly, a recent multi-model study strongly suggests that solar geoengineering cannot simulta-neously return regional and global temperature and hydrologic cycle intensity to preindustrial levels (Kravitz et al. 2014b).

Demands on climate models

Climate models must fulfill two different roles in the context of geoengineer-ing research and possible implementation. The starting point of respective modeling studies is to provide a useful and reliable answer to whether the various proposed climate engineering schemes potentially fulfill the foreseen purpose, in general to substantially cool climate systems to counteract global warming. The second main task is to assess and quantify the occurrence of unintended side effects, like an accompanying reduction of rainfall, the disloca-tion of the monsoon areas or the perturbation of the ozone layer in the case of stratospheric particle injection.

Currently available climate models are useful, but they are also far from "perfect" (Schmidt 2009). Many uncertainties confining the model runs are related to how they deal with small scale processes that slip through the model grid and thus need to be parameterized (Donner and Large 2008; Edwards 2010). These include cloud microphysical processes, which con-trol the life cycle of clouds and the formation of precipitation, which are especially poorly represented in global models. As Hélène Guillemot (in this volume) shows, the development of new cloud parameterizations is a key strategy of domesticating the uncertainty of climate models, albeit with no guarantee of success.

For almost all the schemes proposed, climate engineering in the category of SRM induces an imbalance in the spatial energy distribution, which regionally might lead to compensating flows and related changes in cloud fields and lifetimes. The extent to which heterogeneous radiative forcing patterns can lead to circulation changes, which can, in turn, influence the temperature change pattern and subsequently precipitation, is an under-researched topic. The pattern of temperature change seems to be determined by a combination of the forcing pattern and the feedback processes involving clouds, a subject of high uncertainties in existing climate models (Stephens 2005; Bony et al. 2006). A distorted pattern in the geographical distribution of the ratio of shortwave incoming and long-wave outgoing radiation is expected especially in the case of marine cloud seeding, which can only be implemented meaningfully in certain regions of the world oceans, namely in the large marine stratocumulus fields west of the continents. Associated spatial, temporal, and quantitative changes in precipitation patterns are to be expected but are not modeled with any sufficient certainty, as the results from simulated geoengineering scenarios for the most recent IPCC report strikingly document (Boucher et al. 2013). Concerning changes in precipitation for many regions of the globe, the models did not even agree on the direction of the effect. It should be mentioned that most model studies addressing geoengineering aspects in the field of SRM use a model grid with inadequately low resolution. Model studies looking at the effectiveness and side effects of aerosols injections into the stratosphere require not only a sophisticated stratospheric chemistry module, but also an aerosol model capable of depicting the initial injection phase with respect to the evolution of the relevant particle properties, such as size distribution and number concentration. Neither of these modules is available in sufficient quality in existing standard climate models.

Modeling the effects of CDR may be somewhat easier, since in general the relevant time scales are generally much longer and the system comes closer to equilibrium without large imbalances in radiative forcing patterns. Here the implementation of measures involving natural cycles like afforestation or ocean fertilization seems to be the bottleneck. There is a need for dynamically coupled terrestrial and marine ecosystem modules linked to the biogeochemical cycles in the Earth system models to assess their effectiveness in a changing climate, since, for example, warming and acidification will continue during an implementation program. Today, such ecosystem sub-models are still in their infancy.

Overall, it seems justified to state that climate models to be used for full assessments of geoengineering proposals, including expected and unexpected side effects, need to meet the highest standards.

Instructions: Policy advice and economic utilizability

In the case of policy advice for geoengineering, there is greater demand with respect to the reliability and quality of model results than on those climate model projections that advise global climate politics in general (i.e. the scenario-driven and attribution-related model runs underlying the IPCC assessments). The latter require much less accuracy regarding regionalization and accurate quantification, since their main purpose is to provide a guideline or range for policy options, for which tendencies or corridors of possible outcomes are often sufficient. Climate models cannot currently be employed to pinpoint with certainty what and in which magnitude changes could be expected in different regions on the globe. In contrast, the latter information is of utmost importance before geoengineering measures can be implemented. Affected nations or regions and responsible authorities need to be fully informed in advance about adverse side effects to be expected before geoengineering is seriously considered.

Another factor, which led to the cancelation of the SPICE outdoor experiment, is conflicts of interest (Cressey 2012). A patent application for an apparatus used by the project was submitted before the SPICE project proposal, but not declared to the funding agency as a potential conflict of interest. This is against the rules of UK funding agencies. There is certainly a need to separate research from economic interests because geoengineering is not only a business model but bears the danger of significant interference with the Earth system.

Even more challenges were posed by an experiment performed by the US businessman Russ George. In July 2012, he conducted an iron dumping experiment under the aegis of the Haida Salmon Restoration Corporation (HSRC). HSRC injected 120 tons of ferrous sulfate into the Pacific Ocean, an amount five times larger than in all small-scale experiments to date (Toussaint 2013). The stated aim of this project was to restore the local salmon population by fertilizing the marine biosphere. However, as a side effect, ocean iron fertilization might also reduce atmospheric concentrations of carbon dioxide and thus pave a new way to gain carbon credits. The experiment and the monitoring were conducted without any governmental or scientific control. Such unilateralism represents a significant governance challenge and equally problematic is the blending of environmental and economic interests.

Models as they are employed in present-day geoengineering research cannot reproduce the observed climate in precise detail, for instance regional patterns and temporal distributions of precipitation are not well matched. But it is often this regional scale on which policy advice is sought by decision makers and authorities (see also Mahony in this volume). In climate change research results are typically downscaled, either statistically, or dynamically using regional climate models (RCMs). These approaches both carry their own deficiencies (e.g. cold/warm biases in the case of dynamical downscaling) and have not yet been applied in geoengineering studies.

Another issue related to suitability for policy advice is that models in use today provide results for the basic geoscientific parameters. To be policy relevant, changes in these parameters need to be translated into climate impacts on human systems (risks for human health, economies, the food and water supply) as well as for terrestrial and marine ecosystems (Schäfer et al. 2014). The required impact models in this chain toward policy advice are complex and uncertain and still in a state of infancy.

Conclusion

In the recent past, simulations have played an important role in exploring the potential of geoengineering methods. This is especially the case for studies to explore the SRM methods, whereas CDR methods have also employed in-situ experiments (ocean fertilization by iron oxides). The reasons for this difference are that ocean fertilization is rather cheap and technically much less challenging than SRM experiments, and that modeling tools to simulate the marine biosphere are much less advanced than Earth-atmosphere models. This is because models of the atmosphere system are mainly based on physical laws whereas models of the biological system are based on empirical knowledge, with an inadequate understanding of the relevant processes.

One problem with any intervention in the climate system is the detection and attribution of the effects. The natural variability of the system is high and thus the signal of the intended climate intervention can be obscured by the system's noise. To analyze the effects and the side effects of geoengineering measures, experiments must be performed over many years and the intervention should not be too weak. Therefore, meaningful experiments would come close to actual deployment. Therefore, to date, investigations of SRM methods have been limited to model simulations. If, based on these simulations, recommendations are made to change the climate intentionally, these models should meet higher standards than models used to detect nonintentional anthropogenic climate impacts. Arguably, the stakes are higher for model-based advice in the context of a potential deployment of geoengineering schemes than for climate change projections used to guide international climate policy.

Model simulations have been performed mainly to gain an understanding of the climate system behavior, and in the framework of the IPCC process, to simulate sets of anthropogenic climate change scenarios. But in the years following Paul Crutzen's publication, the number of model simulations investigating the feasibility and possible side effects of geoengineering methods have vastly increased, whereas experiments in the real world are still viewed very skeptically by the public as well as within the scientific community. We might therefore identify an emerging "culture of prediction" in the modeling of geoengineering techniques. Institutionalization is proceeding, for example in the establishment of intercomparison efforts, although it remains to be seen how such knowledge may be used in policy-making, influence wider cultural debates, or seed new practices of domesticating and representing uncertainty. Focused and critical attention is required on the emergence of new epistemic standards in this area, which are defining accuracy and predictive skill in the context of a deeply consequential set of technological proposals. The broader question of the extent to which numerical models, and the possibilities they provide to work through scenarios in a virtual world, facilitate or make more acceptable the idea of geoengineering, or whether the possibility to perform in-silico experiments protects us from real world experiments, is difficult to answer. However, we are living in the "Anthropocene" epoch—a term coined by Paul Crutzen "defined by the dominant human role in the modification of Earth systems" (Minteer 2012). The more obvious it becomes that humans exert a significant impact on climate, the more the position to protect nature's integrity from human manipulation will be challenged (Minteer 2012).

References

Arrhenius, S. (1908) *Worlds in the Making: The Evolution of the Universe*, New York, NY: Harper and Brothers.

Barrett, S. (2008) "The incredible economics of geoengineering," *Environmental and Resource Economics* 39, pp. 45–54.

Bellamy, R., J. Chilvers, N. E. Vaughan and T. M. Lenton (2012) "A review of climate geoengineering appraisals," *WIREs Climate Change* 3(6), pp. 597–615.

Bengtsson, L. (2006) "Geoengineering to confine climate change: Is it at all feasible?" *Climatic Change* 77(3), pp. 229–234.

Bony, S., R. Colman, V. M., Kattsov et al. (2006) "How well do we understand and evaluate climate change feedback processes?" *Journal of Climate* 19, pp. 3445–3482.

Boucher, O., D. Randall, P. Artaxo et al. (eds.) *Climate Change 2013: The Physical Science Basis. Contribution of Working Group I to the Fifth Assessment Report of the Intergovernmental Panel on Climate Change*, Cambridge, MA: Cambridge University Press, pp. 571–657.

Boucher, O., P. M. Forster, N. Gruber, M. Ha-Duong, M. G. Lawrence, T. M. Lenton, A. Maas and N. E. Vaughan (2014) "Rethinking climate geoengineering categorisation in the context of climate change mitigation and adaptation," *WIREs Climate Change* 5, pp. 23–35.

Boyd, P. W. (2008) "Ranking geo-engineering schemes," *Nature Geoscience* 1, pp. 722–724.

Budyko, M. I. (1974). *Izmeniya Klimata*, St. Petersburg, Russia: Gidrometeoizdat.

Budyko, M. I. (1977) *Climatic Changes*, (translation of *Izmeniia Klimata*, 1974) Washington, DC: American Geophysical Union, pp. 236–246.

Cicerone, R. J. (2006) "Encouraging research and overseeing implementation," *Climatic Change* 77(3), pp. 221–226.

Cotton W. R. (2007) "The rise and fall of the science of weather modification by cloud seeding," W. R. Cotton and R. A. Pielke Sr. (eds.) *Human Impacts on Weather and Climate*, Cambridge, MA: Cambridge University Press, pp. 3–8.

Cracken, M. C. (2006) "Geoengineering: Worthy of cautious evaluation?" *Climatic Change* 77(3), pp. 235–243.

Cressey, D. (2012) "Cancelled project spurs debate over geoengineering patents - SPICE research consortium decides not to field-test its technology to reflect the Sun's rays," *Nature* 485, p. 429.

Crutzen, P. J. (2006) "Albedo enhancement by stratospheric sulfur injections: A contribution to resolve a policy dilemma?" *Climatic Change* 77, pp. 211–219.

Donner, L. J. and W. G. Large (2008) "Climate modeling," *Annual Review of Environment and Resources* 33, pp. 1–17.

Edwards, P. N. (2010) *A Vast Machine: Computer Models, Climate Data, and the Politics of Global Warming*, Cambridge, MA: MIT Press.

ENMOD (n.d.) *Prohibition of Military or Any Other Hostile Use of Environmental Modification Techniques*, United Nations Convention, available from: http://www.un-documents. net/enmod.htm, accessed 2 April 2015.

Feichter, J. and T. Leisner (2009) "Climate engineering: a critical review of approaches to modify the global energy balance," *The European Physical Journal Special Topics* 176, pp. 81–92.

Fleming, J. R. (2006) "The pathological history of weather and climate modification: Three cycles of promise and hype," *Historical Studies in the Physical Sciences* 37, pp. 3–25.

Fleming, J. R. (2007) "The Climate Engineers," *Wilson Quarterly* 31, pp. 46–60.

Hansen, J., A. Lacis, R. Ruedy and M. Sako (1992) "Potential climate impact of Mount Pinatubo eruption," *Geophysical Research Letters* 19, pp. 215–218.

Hegerl, G. C. and S. Solomon (2009) "Risks of climate engineering," *Science* 325, pp. 955–956.

House K. Z., A. C. Baclig, M. Ranjan, E. A. van Nierop, J. Wilcox and H. J. Herzog (2011) "An economic and energetic analysis of capturing CO_2 from ambient air," *PNAS Proceedings of the National Academy of Sciences* 108, pp. 20428–20433.

Jones, A., J. Haywood and O. Boucher (2009) "Climate impacts of geoengineering marine stratocumulus clouds," *Journal of Geophysical Research* 114(10), D10106.

Jones, A. and J. M. Haywood (2012) "Sea-spray geoengineering in the HadGEM2-ES earth-system model: radiative impact and climate response," *Atmospheric Chemistry and Physics* 12, pp. 10887–10898.

Keith, D. W. (2000) "Geoengineering the climate: History and prospect," *Annual Review of Engery and the Environment* 25, pp. 245–284.

Kellogg, W. W. and S. H. Schneider (1974) "Climate stabilization: For better or for worse?" *Science* 186, pp. 1163–1172.

Kiehl, J.T. (2006) "Geoengineering climate change: Treating the symptom over the cause. An Editorial Comment," *Climatic Change* 77(3), pp. 227–228.

Kravitz B., K. Caldeira, O. Boucher et al. (2011) "The geoengineering model intercomparison project (GeoMIP)," *Atmospheric Science Letters* 12, pp. 162–167.

Kravitz B., K. Caldeira, O. Boucher et al. (2013) "Climate model response from the Geoengineering Model Intercomparison Project," *Journal of Geophysical Research* 118, pp. 8320–8332.

Kravitz, B., A. Robock and O. Boucher (2014a) "Future directions in simulating solar geoengineering," *Eos* 95(31), p. 280.

Kravitz, B., D. G. MacMartin, A. Robock et al. (2014b) "A multi-model assessment of regional climate disparities caused by solar geoengineering," *Environmental Research Letters* 9, 074013.

Lamb, H.H. (1971) "Climate-engineering schemes to meet a climatic emergency," *Earth-Science Reviews* 7, pp. 87–95.

Lane L., K. Caldeira, R. Chatfield and S. Langhoff (2007) *Workshop Report on Solar Radiation Management*, (Technical Report from workshop held at Ames Research Center on November 18–19, 2006), Washington, DC: National Aeronautics and Space Administration.

Lawrence, M. G. (2006) "The geoengineering dilemma: To speak or not to speak," *Climatic Change* 77(3), pp 245–248.

Lenton, T. M. and N. E. Vaughan (2009) "The radiative forcing potential of different climate geoengineering options," *Atmospheric Chemistry and Physics* 9, pp. 5539–5561.

Marchetti, C. (1977) "On geoengineering and the CO2 problem," *Climatic Change* 1, pp. 59–68.

Minteer, B. A. (2012) "Geoengineering and ecological ethics in the anthropocene," *BioScience* 62(10), pp. 857–858.

NAS National Academy of Sciences (1992) "Chapter 28: Geoengineering," NSA (ed.) *Policy Implications of Greenhouse Warming: Mitigation, Adaptation, and the Science Base*, Washington, DC: National Academies Press, pp. 433–464.

Rasch, P. J., S. Tilmes, R. P. Turco, A. Robock, L. Oman, C. -C. Chen, G. L. Stenchikov and R. R. Garcia (2008) "An overview of geoengineering of climate using stratospheric sulphate aerosols," *Philosophical Transactions of the Royal Society A* 366, pp. 4007–4037.

Rickels, W., G. Klepper, J. Dovern et al. (2011) *Large-Scale Intentional Interventions into the Climate System? Assessing the Climate Engineering Debate*, Kiel, Germany: Kiel Earth Institute.

Royal Society (2009) *Geoengineering the Climate: Science, Governance and Uncertainty*, London: The Royal Society.

Rusin, N. and L. Flit (1960) *Man versus Climate*, Moscow, Russia: Peace Publishers.

Schäfer, S., A. Maas and P. Irvine (2013) "Bridging the gap in interdisciplinary research on solar radiation management," *GAIA - Ecological Perspectives for Science and Society* 22(4), pp. 242–247.

Schellnhuber H. J. (1999) "'Earth system' analysis and the second Copernican revolution," *Nature* 402(Supp.), pp. C19–C23.

Schmidt, G. (2009) "Wrong but useful," *Physics World* 13(10), pp. 33–35.

SPICE Project (2013) Homapage available from http://www.spice.ac.uk/, accessed 20 February 2017

Stephens, G.L. (2005) "Cloud feedbacks in the climate system: A critical review," *Journal of Climate* 18, pp. 237–272.

Teller, E., L. Wood and R. Hyde (1997) *Global Warming and Ice Ages: I. Prospects For Physics-Based Modulation Of Global Change*, (University of California Research Laboratory Report UCRL-JC-128715), Livermore, CA: Lawrence Livermore National Laboratories.

Thereluctantgeoengineer (2012) *Blog*, available from http://thereluctantgeoengineer.blogs-pot.co.uk/ 2012/05/testbed-news.html, accessed 5 May 2012.

Tilmes, S., J. Fasullo, J.-F. Lamarque et al. (2013) "The hydrological impact of geoengi-neering in the Geoengineering Model Intercomparison Project (GeoMIP)," *Journal of Geophysical Research: Atmospheres* 11(118), pp. 11036–11058.

Tilmes, S., M. J. Mills, U. Niemeier et al. (2014) "A new geoengineering model inter-comparison project (GeoMIP) experiment designed for climate and chemistry models," *Geoscientific Model Development* 7, pp. 5447–5464.

Toussaint, P.A. (2013) *Regulation under Scientific Uncertainty: Effectiveness, Impacts and Governance of Ocean Iron Fertilization*, (Master Thesis), Vienna: Environmental Technology and International Affairs.

Robock, A. (2007) *Re: Solar Radiation Management workshop report*, (reply to Stephanie Langhoff, 23 Feb 2007).

UNFCCC (1992) *United Nations Framework Convention on Climate Change*, available from https://unfccc.int/resource/docs/convkp/conveng.pdf, accessed 20 February 2017.

UN United Nations (1972) *Conference on the Human Environment. Declaration*, available from http://www.un-documents.net/unchedec.htm, accessed 2 April 2015.

Vaughan, N. E. and T. M. Lenton (2011) "A review of climate geoengineering proposals," *Climatic Change* 109, pp. 745–790.

Weart, S. P. (2011) "Climate modification schemes," available from http://www.aip.org/history/climate, accessed 15 November 2013.

Wexler, H. (1958) "Modifying weather on a large scale," *Science* 128, pp. 1059–1063.

Williamson, P. (2016) "Emissions reduction. Scrutinize CO2 removal methods," *Nature* 530(7589), pp. 153–155.

11 Validating models in the face of uncertainty

Geotechnical engineering and dike vulnerability in the Netherlands

Matthijs Kouw

Introduction

The geographical position of the Netherlands makes it crucial to assess the safety of Dutch flood defenses. About 3.4 million Dutch (21 percent of the total population) live below sea level (Centraal Bureau voor de Statistiek 2008: 65). Nineteen percent of the gross national product (GNP) is earned below sea level, although a total of 32 percent of GNP is earned in areas that are prone to flooding (Centraal Bureau voor de Statistiek 2008: 64). Geotechnical engineering, a subdiscipline of civil engineering concerned with the behavior of soil under different conditions, fulfills a crucial function in this regard by modeling processes that cause flood defenses (e.g. dikes, dams, and sluices) to fail. Such processes are also known as failure mechanisms. Geotechnical modeling relies heavily on both physical models (e.g. scale models of flood defenses that are subjected to water pressures) and computational models (e.g. calculation rules that simulate the relationships between soil morphology and structural stability). Geotechnical models need to be validated to determine their ability to provide an accurate and reliable assessment of the safety of flood defenses.

As various studies of modeling practices have shown, pragmatic and contextual considerations shape model validation (e.g. Morgan and Morrison 1999; Oreskes et al. 1994; Winsberg 2006). Morgan and Morrison (1999) argue that modeling should not be interpreted exclusively in terms of mirroring or mimesis of target systems, but also with close attention to the demands of particular settings: "[W]e do not assess each model based on its ability to accurately mirror a system, rather the legitimacy of each different representation is a function of the model's performance in specific contexts" (Morgan and Morrison 1999: 28). Thus, the performance of models can be assessed in terms of "relevance" rather than truth. In this perspective, social groups attribute explanatory power and reliability to models by virtue of the latter's contribution to solve a particular problem, making the models in question relevant for these social groups.

Drawing on insights from Science and Technology Studies (STS) and an ethnographic study of geotechnical engineering conducted at Deltares (a Dutch institute for applied research on water, subsurface and infrastructure) between 2009 and 2011, this chapter examines modeling practices and the validation of models pertaining to research on a dike failure mechanism known as "piping," which is a form of seepage erosion. Calculation rules and models of piping serve to predict the risk of dike failure. Piping research and modeling may be regarded a specific case of a culture of prediction in geotechnical engineering. As will become clear, research on piping features a series of steps and model-related forms of knowledge production, where each step produces knowledge that is made available for subsequent steps. Of particular importance are the development and adoption of computational models.

Over the course of the twentieth century, geotechnical engineering has come to rely more heavily on computational models (i.e. models based on mathematical insights that require computational resources to run simulations of complex systems). This trend can be attributed to water management across the board (Kouw 2016). Disco and van den Ende (2003) explain the widespread adoption of computational models by pointing out that such models fulfilled a crucial role as management tools in Dutch water management, and met a more general desire to quantify water-related phenomena. The successful application of computational models implies "black-boxing" (Latour 1987, 1999): "When a machine runs efficiently […] one need focus only on its inputs and outputs and not on its internal complexity. Thus, paradoxically, the more science and technology succeed, the more opaque and obscure they become" (Latour 1999: 304). The successful application of black-boxed technologies, in this case computational models, means they are taken for granted and only come into view when failure or malfunctioning renders them obtrusive.

Uncertainty features prominently in all steps of the modeling chain deployed in the case of piping. Uncertainty is sometimes defined as a lack of knowledge (Petersen 2012; Kouw et al. 2013). I adopt Gross' (2010) definition of uncertainty as "a situation in which, given current knowledge, there are multiple possible future outcomes" (Gross 2010: 3). Uncertainty can produce new insights about risks: "multiple possible future outcomes" might produce insights about risks and what to do about them. Various forms of uncertainty emerge in geotechnical modeling, and social groups deal with these uncertainties in diverging ways. The use of geotechnical models in the laboratory can serve to investigate uncertainties of geotechnical phenomena and to acquire a deeper understanding of these phenomena. Outside of the laboratory, users of geotechnical models may be less inclined to study the uncertainties of geotechnical phenomena. In this regard, black-boxed technologies can travel easily from the laboratory to contexts outside of the laboratory (e.g. decision making and policy making). When accepted without further questioning, black-boxed geotechnical models may cause users of such models to gloss over uncertainties. Black-boxing geotechnical models, hence, is a powerful way of domesticating uncertainty and making it largely invisible to its users. This paper shows how black-boxing occurs in various steps of piping-related modeling, and argues that black-boxing may not bode well for the potential of uncertainty to function as a source of knowledge, which may negatively impact the safety of the Netherlands.

The main questions of this chapter are as follows: how do geotechnical models contribute to the production of knowledge about dike failure mechanisms that is considered relevant by the social groups involved, and how may the various ways in which these social groups deal with the uncertainties involved with the use of geotechnical models put the Netherlands at risk? I address these questions by first describing how geotechnical engineers deploy modeling in their study of piping. Subsequently, I describe how knowledge thus developed is used in social domains outside of geotechnical engineering. In both cases I address the black-boxing of knowledge, what knowledge is considered relevant for the social groups involved, and how uncertainties that arise are addressed.

Piping

Piping is a form of seepage erosion involving the movement of water under or through a dike that provokes instability, in some cases leading to dike breaches and even dike failure. High water levels lead to high water pressure or "hydraulic head" on the water side of the dike, which may cause a flow of water under or through a dike. This flow can build channels or "pipes," which eventually form a shortcut between the two sides of the dike and run through the dike and/or its foundations. In such cases, water wells up through soil (also known as a "sand boil"), which is an important visual indication that piping is in progress. Shortcuts between the dike's water and land sides transport large amounts of soil and dramatically increase the speed of erosion, which may damage the dike or its foundations to such an extent that the dike collapses or breaches. In the Netherlands, many dikes consist of clay and/or peat that sit on foundations of sand, particularly in the vicinity of the main rivers of the Netherlands. Since clay and peat are cohesive and relatively impermeable while sand is relatively permeable, many dikes in the Netherlands are prone to seepage erosion of their foundations.

The composition of dikes and their foundations, and the interactions between different types of soil in dikes and their foundations, are sources of uncertainty in geotechnical engineering. The composition of soil may be known at locations where measurements have been taken, but soil can be rather heterogeneous, implying major differences between measuring points. In addition, geotechnical engineers stress the difficulties imposed by the complexity of interactions between different kinds of soil. Such interactions are not understood very well yet, and remain a source of uncertainty.

To gain an understanding of the behavior of soil, geotechnical engineers rely heavily on experiential knowledge. There are only a few detailed observational accounts of the piping process. More importantly, most of the piping process is inaccessible to the human senses, since it takes place inside a dike. Today, physical and computational models provide important extensions of the human senses, allowing geotechnical engineers to study phenomena otherwise inaccessible to them. Physical models of dike foundations on different scales provide the means to study the conditions that provoke piping, how piping proceeds, and what conditions influence the onset and progress of piping, for example, the composition of the dike's foundations and the hydraulic head.

Differences in the shape and size of grains of sand make for different types of sand, which also behave differently under pressure. To acquire an understanding of piping, qualitative physical experiments are carried out using a cross section of the foundations of a hypothetical dike. A Plexiglas window covers the cross section so that the process of piping can be observed. Water pressure is applied on one side of the cross section to simulate the hydraulic head that provokes the onset of piping. A part of the cross section is covered with a counterweight to simulate the pressure exerted by the top layer of the dike. Part of the cross section on the right-hand side is left open to simulate the presence of a ditch, which can offer a way for the water to come to the surface due to the water pressure exerted by the hydraulic head.

Based on empirical observations acquired during physical experiments, calculation rules can be devised and validated. An example of such calculation rules can be found already in the early twentieth century, when the British Colonel Bligh concluded that the loss of hydraulic head is proportional to the distance water travels (also known as creep length). Increasing creep length can be an important way to decrease the risk of seepage erosion (Bligh 1910). Similarly, calculation rules pertaining to piping describe relationships between hydraulic head, soil properties, and creep length.

Calculation rules are needed to develop computational models of piping and once formalization in the form of calculation rules is possible, it is possible to develop computational models that run simulations based on these calculation rules. In this regard, it is possible in principle to develop quantitative approaches to geotechnical phenomena. Formalization in the form of calculation rules in combination with quantitative measurement of certain phenomena relevant to piping (e.g. hydraulic head, creep length) allows the risk of piping to be predicted. In the following, I refer to this combination of calculation rules and measurement as quantitative methods. However, calculation rules currently do not fully describe and predict piping, making it necessary to introduce empirical parameters based on physical experiments.

Earlier physical experiments in the 1990s were not carried out to the point where a "full" pipe acted as a shortcut between the water and land side of the dike, since this would have damaged the experimental setup (Vrijling et al. 2010: 41). As a result, the hydraulic head that would provoke "retrograde erosion," where a pipe forms a shortcut between the dike's water and land sides, was not determined. A further shortcoming of these earlier experiments on piping is that the highly influential morphological properties of soil were not studied exhaustively. The critical head is influenced by the thickness of the sand layer and top layer in question, the permeability of the sand layer, and soil morphology (e.g. size and shape of sand grains). Despite these interacting complexities, initial calculation rules developed to calculate critical head assumed the homogeneity of soil.

Piping found its way back to the research agenda of Deltares in 2007. An important influence in this was the *Veiligheid Nederland in Kaart* (VNK or Mapping the Safety of the Netherlands) effort, a collaboration between the

Dutch Ministry of Infrastructure and the Environment, the water boards of the Netherlands,[1] and the Interprovinciaal Overleg (a foundation comprising the provinces of the Netherlands as members). The first phase of VNK took place between 2001 and 2005, and concluded that piping posed a substantial risk to dike safety in the Netherlands (Rijkswaterstaat 2005: 90). The calculations used in VNK are based on scenarios that include extreme water levels that have never been observed. However, these hypothetical water levels had very concrete repercussions. During the first phase of VNK, the shortcomings of calculation rules developed in the 1990s became the subject of debate (Vrijling et al. 2004). When the use of these calculation rules led to high estimations of dike failure due to piping, the various parties involved with VNK found it necessary to improve the accuracy and reliability of these calculation rules. As a result, a new round of research on piping commenced in 2007.

Small, medium, and full-scale physical experiments

Experiments similar to those in the 1990s were carried out using small-scale (see Figure 11.1) and medium-scale physical models. An important motivation behind these experiments was the desire to acquire observational knowledge of the piping process.

When I attended a physical experiment using a medium-scale physical model, I was introduced to some of the challenges related to the study of piping. During the experiment, the model was covered with a thick sheet of black plastic to keep sunlight out and minimize reflections on the Plexiglas sheet that covered the layer of sand that was studied. Light and reflections can compromise the quality of the camera recordings used to capture the process of piping. The lamps used to illuminate the experiment generated heat, introducing discomfort on the part of the scientists, for whom tracing the movements of individual grains of sand required utmost concentration. More than once, a moving grain of sand was a source of modest celebration or at least a welcome change in an otherwise fairly uneventful experiment. The experimenters concentrated on the movement of individual particles, and studied how the meandering flows of water created small channels that would sometimes persist, but could also disappear quickly. When the experiment was not very eventful, the water pressure that simulated the hydraulic head would be increased. This was usually not done according to an exact and elaborate protocol, but rather to provoke some kind of worthwhile event, for example, moving grains of sand or the buildup of meandering channels.

The use of geotechnical models on a scale smaller than the target systems in question leads to different behavior of soil (e.g. due to different effects of gravity). As a result, some phenomena observed in a physical model may occur only in the laboratory, and may therefore not be representative of their target systems. To provide more elaborate means of studying piping and the calibration and validation of geotechnical models, geotechnical engineers have

Figure 11.1 Physical model used for small-scale piping experiments. Water is forced to flow from the bucket on the right hand side through the cross-section of a dike underneath the Plexiglas window. Increasing the bucket's height simulates a greater level of hydraulic head.

Source: Photo by Vera van Beek, courtesy of Deltares.

conducted physical experiments using full-scale dikes as part of the so-called 'IJkdijk' (literally 'calibration dike') program (see Figure 11.2).

The IJkdijk experiments provided additional insights into the onset and progress of piping. A total of four experiments related to piping were conducted at the time of writing, in all cases leading to dike failure, proving once and for all that piping needs to be taken seriously. According to the engineer leading Deltares' piping research, this result was expected by the geotechnical engineers involved. However, several water boards were not convinced piping was really an issue until the IJkdijk experiments (interview, 27 May 2009). The IJkdijk program, therefore, had an important persuasive effect as well.

Figure 11.2 An IJkdijk after an experiment in late 2009.

Source: Photo by Vera van Beek, courtesy of Deltares.

The ensemble of small, medium, and large-scale physical models used by geotechnical engineers helped to address the complexities of soil morphologies and the uncertainties introduced by modeling geotechnical phenomena on different scales. When IJkdijk experiments validate the outcomes of smaller-scale physical models, the latter are considered more reliable. This reduces the necessity of conducting expensive physical experiments on a full scale, and allows smaller-scale experiments to be conducted with more confidence. An important outcome of the small and medium-scale physical models in combination with the IJkdijk experiments was the correction of existing calculation rules. These calculation rules did not predict critical head correctly in the case of coarse sand particles. A further result of the IJkdijk experiments was that geotechnical engineers learned more about the time it takes for a dike to fail because of piping. For example, retrograde erosion turned out to take much longer than expected. Small-scale physical models usually showed a single channel where the process of retrograde erosion proceeded quickly. In large-scale physical models, the process of retrograde erosion could occur suddenly and violently, but could also take several days.

Relevant knowledge and uncertainties in geotechnical research on piping

Although much progress has been made in terms of validating existing calculation rules used to assess piping-related risks, the research is not complete.

In fact, geotechnical engineers question the ability of calculation rules to provide robust predictions, as not all aspects of piping are understood and represented sufficiently. One geotechnical engineer involved with the modeling of piping put it as follows:

> A risk with large-scale physical experiments is that you try to validate too many things, and that is not possible. So the setup has been relatively simple, as were the aims of the model validation. But you cannot validate all of the aspects of the model. Eventually you will get a critical head in the form of a number, and the only thing you can do is check whether that number corresponds with what we thought, and yes, on that basis you need to trust the model, but you cannot validate all aspects. That is tricky. That requires many more experiments.
>
> (interview, 24 June 2011)

Despite these shortcomings, Rijkswaterstaat (an organization that is part of the Dutch Ministry of Infrastructure and the Environment and is responsible for designing, constructing, managing, and maintaining the main infrastructure facilities in the Netherlands) considered the process of validation that was carried out as sufficiently thorough. However, the head of piping research at Deltares points out that additional physical experiments are needed as a basis for comparing the outcomes of different runs of computational models:

> The research does not rid you of the problem of deducing simple calculation rules. You keep discovering new blind spots. The moment you have a calculation rule, it may be state of the art, but that does not mean you are really at the end of the research [...] one experiment is no experiment, you always need to compare the results of different experiments, but it is always a question of time and money [...] *Rijkswaterstaat* expects us to come with a new calculation rule this year, so there comes a point where you have to say good is good enough. But uncertainties remain.
>
> (interview, 27 May 2009)

The head of piping research at Deltares further explains that the outcome of piping-related modeling can be counterproductive in terms of reducing uncertainties in calculation rules. More knowledge about piping can also lead to the realization that more uncertainties apply to piping, which may unsettle the credibility of calculation rules previously deemed trustworthy. For example, the shape of sand grains may turn out to be a complicating factor, which would give rise to the need to incorporate details on sand granularity in calculation rules. As a result, the head of piping research at Deltares argues, it may be unlikely that geotechnical phenomena can be captured once and for all in calculation rules due to the complexities pertaining to such phenomena (interview, 27 May 2009).

Another geotechnical engineer at Deltares working on piping expressed his doubts about attempts to capture piping once and for all in a calculation rule:

> I do not believe in a calculation rule that represents reality. The phenomenon features lots of different aspects, and you can never capture those correctly. You have to provide a schematization of reality before you can start calculating, and reality is so complicated. Those sand layers can be one centimeter thick, they can be small, large, vertical, and horizontal, making the soil so heterogeneous you cannot capture it in a single calculation rule.
>
> (interview, 26 May 2009)

The physical experiments in the laboratory provide ample evidence for this particular engineer's observation that piping is a rather complex and local phenomenon, in which the interactions of heterogeneous soil can have a crucial effect. In principle, vast quantities of information about soil could make a difference, but measuring soil in great detail introduces practical limitations (e.g. available resources; accessibility of measuring points). In addition, the onset and process of piping can be sudden, making even the hypothetical scenario of perfect computational models in combination with exhaustive data about soil problematic in terms of preventing piping altogether.

Other difficulties are related to experimental setups, which may introduce additional uncertainties. For example, geotechnical engineers need to find out what types of sand need to be used in the cross section of physical models, ensure the water pressures used correspond to the conditions of dikes in the Netherlands, and determine whether the Plexiglas cover exerts the right pressure on the model foundation. In the case of the IJkdijk program, producers of measuring devices and sensors were eager to fill the dikes used during IJkdijk experiments with measuring devices, which came to a point where the devices could influence the experiment, as they were located on the border between the sand layer and the clay of the dike. Further complications arose due to the use of generators near the area where the IJkdijk experiments took place—a remote site in the north of the Netherlands. These generators provided power necessary for lamps and other devices, but may also have influenced the experiment by generating vibrations that introduce noise measurements. However, it is not uncommon for such vibrations to occur in the case of a "real" dike whenever trucks or ships pass by.

By means of elaborate simulations with an ensemble of physical models, existing calculation rules used to assess piping-related risks are validated and improved where necessary. The use of geotechnical models in research on piping revealed sources of uncertainty that warrant further research, for example, the difficulties encountered in the laboratory, such as the challenges of understanding soil morphologies and difficulties associated with the experimental setting of geotechnical models. In addition, the issue of scaling may reduce the reliability of small- and medium-scale physical models. These aspects of piping need to be addressed by future research, which is dependent on the allocation of resources from parties like Rijkswaterstaat or companies that consider projects like the IJkdijk to be worthwhile. Geotechnical models need to perform within the specificities of geotechnical research relevant for dike safety policies by producing "deliverables," in this case state of the art calculation rules that are considered to be reliable not only by geotechnical engineers, but also by other social groups, including decision makers, policy makers, and stakeholders. In the following, I show how calculation rules become black-boxed in the form of software despite the previously mentioned uncertainties. This does not bode well for the ability of social groups outside the domain of geotechnical engineering to grasp the full scope and impact of uncertainties that arise with the use of geotechnical models to study piping.

From experimentation to data gathering

Within geotechnical engineering, models fulfill a primarily heuristic role by virtue of being representations "useful for guiding further study but not susceptible to proof" (Oreskes et al. 1994: 644). In other social domains that make use of geotechnical models, the latter fulfill the role of representations used for flood risk management, safety assessments, and dike safety policies. Thus, the role of geotechnical models cannot be framed exclusively in terms of their exploratory function. In this section, I elaborate on the representational role of geotechnical models in social domains outside of geotechnical engineering. The quantitative methods described in the previous section referred to calculation rules developed on the basis of empirical observations. Although such quantitative methods return in this section, I refer to "data-intensive" methods where I discuss quantitative methods that are augmented by large amounts of data and computational power. In this context, quantitative methods produce the perception that geotechnical models provide reliable explanations.

As I showed in more detail in the previous section, geotechnical engineers do not consider the availability of calculation rules as indicative of a complete understanding of piping. Still, physical models are often abandoned in favor of computational models. As the head of research on piping at Deltares put it:

> As soon as there is a degree of certainty about the process, physical models are supposedly no longer needed [...] I think you still need to look at physical models to get some kind of sense of phenomena. If you only work with computational models you might distance yourself too much from reality.
>
> (interview, 27 May 2009)

According to all the geotechnical engineers at Deltares I interviewed, validated geotechnical models are not really "true" in a literal sense. Rather, validated geotechnical models are true in a pragmatic sense and can be considered sufficiently reliable in terms of understanding and predicting piping. Many of the engineers display a belief in "progressive understanding": At some point, calculation rules about piping are considered to be reliable, allowing the codification of knowledge in the form of a calculation rule, which allow quantitative methods that signal a departure from qualitative physical experiments in the laboratory.

The viability of data-intensive methods is based on the presumption of computational tractability—the ability to quantify phenomena and subsequently predict or monitor these phenomena using computational methods. However, social groups have differing commitments to computational tractability. Two engineers working at the Netherlands Organisation for Applied Scientific Research (TNO) that I interviewed argued that geotechnical modeling and data-intensive methods can be combined to create a novel approach to dike safety. For example, data about past events can be fed into a database, which can then be consulted to predict the likely behavior of a dike in those cases where present circumstances are similar to those in the past. One of the engineers explained this as follows:

> You do not have to understand geotechnical phenomena to be able to predict them [...] if you can analyze a large amount of data by means of Artificial Intelligence, you can make statements about the future without understanding the process [...] a dike watcher will do the very same on the basis of past experiences and common sense without having a clue about what goes on inside the dike.
>
> (interview, 30 July 2009)

Although dependent on the acquisition of data and the accuracy of data collected, data-intensive techniques can guide the attention of experts and can point out which dikes need to be subjected to further scrutiny, for example, by carrying out structural improvements or monitoring their status more closely. Calculation rules may pave the road for quantitative approaches that shift the focus of engineers away from physical experimentation, and justify an emphasis on monitoring techniques that focus on data generation and data management. Presentations on the value of monitoring techniques are usually combined with references to "innovative" technologies, such as laser imaging detection and ranging (LIDAR), which is used to detect dents in the surface of dikes that can indicate damage in its structural integrity; remote sensing, which can detect temperature differences that can indicate the permeation of water in a dike that might be caused by damage inside the dike; and the use of sensors to monitor temperature and humidity.

Despite these promising developments, quantitative techniques should be approached with caution. When measuring devices are too far apart, a pipe can simply disappear "under the radar" and remain unnoticed. A further problem is that it is unclear how long it takes for a dike to fail as a result of piping. Although computational models allow sophisticated calculations, they can also be used without understanding the underlying processes and the availability of sufficient data to validate the model in question. The complexity of soil morphology and the lack of data about soil problematize the validation of computational models. A strong reliance on such models can lead to wrong assessments, especially in the absence of data to validate the model. A university professor working at the Department of Earth Systems Analysis (ESA) at Technical University Twente further clarifies this potential problem:

> As long as you keep the shortcomings of models in mind, it is fine to rely on computational models. However, when the output of a geotechnical model is used in large-scale projects, things can go awfully wrong. Model output is often accepted as being holy without being subjected to further attention [...] if model output does not differ too much from reality, people simply carry on using computational models.
>
> (interview, 5 June 2009)

It is crucial that the inner workings of the model in question are understood, the university professor quoted above argues in more detail, because the process of validation may only generate more uncertainties. Understanding how computational models yield a particular result enables a degree of control, which can be used to critically assess their output.

Still, geotechnical engineers need to meet the demands of professional environments and the political arena, which often require them to produce quantitative knowledge. Expert judgments are no longer seen merely as a sufficient basis for making decisions in those environments, since they are not unanimously accepted and cannot be controlled easily. The use of data-intensive techniques provides Dutch water management with an innovative edge, and may seem to enable reliable approaches to flood risk management in the eyes of policy makers. In sum, there are different and not necessarily compatible commitments to the idea of computational tractability. Geotechnical engineers tend to interpret the output of physical and computational models as a result that needs to be revised constantly in the light of new research results. In the eyes of members of other social groups, such as decision makers, policymakers, and stakeholders, computational tractability is more likely to enable monitoring techniques that are valued as reliable, innovative, and cutting-edge.

Flood Control 2015

Codified calculation rules enable the dissemination and reproduction of geotechnical knowledge, which can travel outside of the laboratory to policy contexts in the form of software applications. The use of data-intensive methods further adds to the perceived credibility of such applications. Flood Control 2015, a consortium made up of commercial companies and governmental institutions (i.e. Arcadis, Deltares, Fugro, Royal Haskoning, HKV, IBM, ITC, Stichting IJkdijk, and TNO), aims to develop data-intensive applications for flood mitigation. These applications are in many cases aimed at decision makers, policy makers, and stakeholders, and are used for measuring, monitoring, forecasting, mitigation, and training. More generally, flood risk management increasingly embraces the process of translating expert knowledge to the operational contexts of decision makers, policy makers, and stakeholders, which is expected to lead to robust and participatory forms of flood risk management. The consortium produced a series of applications that raised significant interest in the world of flood risk management. These applications have been praised as innovative and cutting-edge technologies that translate geotechnical knowledge to a public of non-experts. The Flood Control 2015 project is emblematic of the shift to a more adaptive style of flood risk management, since it contains many projects that display a strong commitment to evacuation and the idea of "preparedness," which "proposes a mode of ordering the future that embraces uncertainty and 'imagines the unimaginable' rather than 'taming' dangerous irruptions through statistical probabilities" (Aradau 2010: 3). Forms of flood risk management that emphasize preparedness imply a new form of citizenship, in which commitments to self-sufficiency shift the responsibility of responding to critical events to citizens.

On 20 January 2010, the Flood Control 2015 consortium organized a symposium that functioned as a showcase of their various projects. Throughout the symposium, the free circulation of accurate information was stressed as a crucial component of successful adaptive strategies. The keynote lecture of the event featured a slide showing a conference room, dubbed the "war room," filled with men (with one single exception), laptops, and beamers projecting maps of the Netherlands and feeds of data related to flood risk and dike safety. Such war rooms can function as central nodes in networks of information that are of crucial important during a crisis, and allow water boards to successfully plan and execute the evacuation of a particular area. "In such situations," the lecturer pointed out, "it is quite pleasant when those present are primarily experts and not politicians." Laughter erupted from the room. Still, the need to bridge the gap between "experts" and "non-experts" was stressed again and again.

The idea of sharing information reverberated throughout the day, but was certainly not embraced unconditionally. A project that bears close semblance to the "war room" environments presented in the keynote lecture is the so-called "Demonstrator Flood Control Room" (DFCR), an interactive user environment that features a variety of applications that can be used to analyze and visualize data from flood and dike monitoring networks. The DFCR functions like a central control platform by integrating data feeds generated by other components of the Flood Control 2015 project, including the sensor networks and remote sensing technologies discussed earlier, which allow it to present weather conditions, water levels, and the status of dikes in a particular area. Although computational models are an important component of the DFCR, running those models often requires a tremendous amount of computational resources and therefore cannot always be applied in crisis scenarios. One possible remedy is to lower the resolution of computational models, dramatically decreasing the time needed to run them. Another solution is to run computational models beforehand using input data that corresponds with scenarios that have a high probability, and subsequently include the output of these model runs in the DFCR. In that case, calculations are not carried out during the actual use of the DFCR, making users reliant on model output rendered before the event of an actual critical situation. An additional use of the DFCR is as a training environment, since it can simulate different scenarios to which users need to respond.

Although participants of the symposium valued the DFCR as a platform to integrate information, its possible implementation was approached with caution. Using the DFCR as a central platform to share data among different parties might make the dissemination of data more efficient and reliable. However, the successful implementation of the DFCR depends on a process of standardization that that is problematic, since local requirements differ from the standards used in the DFCR. The discussion around standardization deals with such practical problems, but also turns to potential dangers—what if uncertainties and assumptions are hidden in the data, which reveal themselves only when it is too late? Black-boxed quantified information can travel more easily to different domains of use in principle, but does not appear to roam about freely. The solutions pertaining to the dissemination of information thus occasionally tend to emphasize technological possibilities rather than considerations related to actual applications.

A related example of applications used to disseminate knowledge from "experts" to "non-experts" is the game called "Levee Patroller" (see Figure 11.3), which was created by a team of software engineers at Deltares who specialize in the development of "serious games"—computer game environments developed for educational purposes.

The Levee Patroller game is currently used to train dike watchers and includes a representation of piping. The game deals with piping by including animations of sand boils (described in section "Piping") to address this failure

Figure 11.3 Screenshot from the "Levee Patroller" game showing a damaged dike.

Source: Courtesy of Deltares.

mechanism. The Levee Patroller emphasizes "procedural skills" rather than the "conceptual understanding" of piping on the part of its users (Harteveld 2011: 233). Players of the Levee Patroller game earn rewards by correctly identifying risks and subsequently reporting those risks to a water management authority. This may be a suitable way to make users of the Levee Patroller aware of the piping phenomenon in general. However, although sand boils indicate that piping is indeed in progress, they do not provide a clear indication of how much the process of retrograde erosion has advanced. What is more, the onset and process of piping can be both gradual and sudden. Once a sand boil is visible, one may already be too late.

A complication related to the dissemination and application of information is that expert knowledge from engineering environments needs to be translated to meet the demands of decision makers, policy makers, and stakeholders. Applications that fit these demands need to be designed, and imply both an enabling and constraining effect on the user's interactions (Akrich 1987). This requires an elaborate process of distilling large amounts of expert knowledge in such a manner that decision makers, policy makers, and stakeholders are presented with information that is considered to be sufficiently detailed for the issues they face in a time of crisis. However, underlying geotechnical models are effectively black-boxed and the technologies in question are presented as innovative and cutting-edge platforms to represent information gathered by means of data-intensive methods.

Organizational challenges also apply to flood risk management. During a session at the Flood Control 2015 event described earlier, participants were asked to enact an evacuation scenario. The session's organizers attempted to tackle the issues that come up during evacuations, especially in the negotiations between local authorities, such as decision makers, the police, and firefighters. The participants discussed whether a single actor should have a mandate that allows him or her to make swift decisions, and how the behavior of citizens and decision makers can be uncertain in times of crisis. Citizens may simply not respond to the request to leave their homes, and decision makers may not decide purely on the basis of information about a critical scenario, which is often already uncertain itself. Although evacuation plans and training for evacuation scenarios were seen in a positive light, participants also stressed the importance of deviating from such plans when necessary.

Relevant knowledge and uncertainties in data-intensive methods and the Flood Control 2015 project

The use of data-intensive methods not only opens up new ways of engaging geotechnical phenomena for engineers, but also facilitates the development of "smart" and "innovative" applications in the form of software, which are expected to enable adaptive forms of flood risk management. Data-intensive techniques provide an important platform for geotechnical engineers to secure resources for further research. Geotechnical engineers can mobilize more resources for doing fundamental research when they also adopt strategies that align well with the Flood Control 2015 program. Thus, the work of the "engineer-entrepreneur" can be analyzed using a "front stage" and "back stage" analogy (Hilgartner 2000; Bijker et al. 2009)—: as much as geotechnical engineers stress the need for fundamental research, their ability to actually do that research depends in part on their ability to position themselves in the framework of innovative flood risk management.

The uncertainties pertaining to quantitative methods and data-intensive methods relate to the questions as to whether such methods suffice, and how quantitative research should be carried out. The process of codification effectively black-boxes geotechnical knowledge in the form of a calculation rule or computer code. This can make it more difficult for users to assess the impact of such calculation rules or computer code. Similarly, the discussion on Flood Control 2015 revealed the use of standardized data, neglecting different local conditions. In addition, the design of applications for decision makers, policy makers, and stakeholders indicated further challenges. Knowledge generated by means of elaborate geotechnical models needs to be made accessible to an audience of non-specialists and fitted to the requirements of flood risk management in action. Standardized data may not be compatible with local contexts and conceal problems. A different source of uncertainties became apparent during the discussion on organizational aspects of decision making in a time of crisis, which looked at the influence of the political interests of decision makers and at idiosyncratic local populations who often act according to their own ideas about risks, making their actions less amenable to control.[2]

Conclusion: uncertainty as a source of innovation?

Geotechnical engineers at Deltares are committed to an elaborate process of research to reduce the uncertainties of geotechnical models, and develop state of the art calculation rules that can be used in safety assessments. The value of geotechnical models is based on their success in specific contexts, which emphasizes relevance rather than truth. In practice, relevance may be confused with truth, particularly outside of the laboratory, where dike safety assessments need to be perceived as epistemically up to par in ways that tie in with organizational, institutional, and political requirements. The use of data-intensive methods indicates a commitment to "innovative" quantitative approaches, and the development of software that fosters preparedness. Geotechnical engineering needs to perform not only according to criteria of epistemic robustness (e.g. by producing more accurate calculation rules), but increasingly also needs to meet demands related to "social robustness" (Nowotny 2003). For geotechnical engineers, this can imply the need to become "engineer-entrepreneurs" (Daston and Galison 2007: 398). Engineers need to produce knowledge that is considered to be relevant by their peer community of geotechnical engineers, but also encounter political commitments to flood mitigation in their work. In this sense, geotechnical engineers "back stage" stress the exploratory capacities of geotechnical models and their limitations. However, "front stage" presentations of such models emphasize representation.

The use of geotechnical models implies a range of uncertainties. Engineers may speak of empirical forms of uncertainty due to the lack of empirical data about soil composition, the complexity of soil behavior, and ensuring the representativeness of physical experiments in the laboratory. The use of geo-technical models outside the laboratory introduces further indeterminacy in the form of organizational challenges in contexts of use. Attempts to develop definitive calculation rules and implement "innovative" technologies for flood risk management can be undermined by uncertainty, defined along the lines of Gross' work as "a situation in which, given current knowledge, there are multiple possible future outcomes" (Gross 2010: 3). Uncertainty can put our highly technological cultures at risk, since the methods chosen to cope with various risks may be out of step with the "multiple possible future outcomes" (ibid.) concomitant with uncertainties. From the perspective of geotechnical engineers, claims to knowledge need to be approached with apprehension— the reliability of geotechnical models does not imply an objective truth, and calculation rules acquire credibility through successful application, which does not mean geotechnical models are complete. However, the black-boxing of knowledge about geotechnical phenomena (e.g. in the form of calculation rules or software) may enable the use of geotechnical knowledge in domains outside of geotechnical engineering where geotechnical models are valued differently.

As became clear, geotechnical modeling may not only contribute to the reduction of uncertainties, but can also lead to awareness of previously veiled uncertainties. Thus, geotechnical modeling may have a disruptive effect in the realm of policy making, since it can lead to new insights about dike failure mechanisms. However, social groups differ in how they value uncertainties (Mackenzie 1999). The settling of knowledge in the form of calculation rules, software, or policies that are considered to be epistemically and/or socially robust is exactly what may put technological cultures at risk. Black-boxed knowledge can imply a diminished ability to evaluate the pros and cons of various approaches to uncertainties, and can preclude the adoption of uncer-tainties as a source of knowledge about risks. Adopting uncertainty as a source of knowledge involves organizational, institutional, and socio-economic chal-lenges. As much as the tractability of geotechnical phenomena can be ques-tioned and the uncertainties involved with geotechnical modeling can be emphasized, it may not be in the interest of social groups to do so. In addition, phenomena of which societies are ignorant cannot always be quantified and turned into probabilities, since they fall outside of the scope of quantitative practices in technological cultures.

Rather than taking a "wait and see" or "wait-until-more-science-is-available" approach to uncertainties,[3] Gross argues that "surprises" need to be deliberately fostered and appreciated as moments where the precarity of objective knowledge becomes apparent. As a result, social groups can become aware of ignorance, identified as "knowledge about the limits of knowing in a certain area," which "increases with every state of new knowledge" (Gross 2010: 68). Surprises can reveal limits of knowledge, and thereby make social groups aware of phenomena that fall outside of existing modes of knowledge production. The acquisition of knowledge can also reveal ignorance. Social experimentation does not aim to "overcome or control unknowns but to live and blossom with them" (Gross 2010: 34). A failure to recognize the value of uncertainty and ignorance as sources of knowledge can put technological cultures at risk. However, uncertainty as a source of knowledge may be kept at bay as a tantalizing promise that turns out to be difficult to realize in practice—uncertainty and ignorance may simply be usurped by vested interests. Even though social experiments need to face vested interests, uncertainty can act as a promising source of innovative knowledge that enhances the resilience of vulnerable societies.

Notes

1   The water boards are regional authorities in charge of the maintenance of flood defenses, waterways, water quality, and sewage treatment. There are currently twenty-five water boards in the Netherlands. The history of the water boards goes back to the thirteenth century, when they developed an elaborate scheme of taxes and governance structures. The water boards are credited as being the oldest form of democratic governance in the Netherlands.
2   Wynne (1992: 117) mentions "indeterminacy" that results from "real open-endedness in the sense that outcomes depend on how intermediate actors will behave." The various applications related to the Flood Control 2015 program feature indeterminacy in the sense that their functioning and value in evacuation procedures will, at least in part, depend on organizational and human components.
3   Joshua Howe calls this approach the "science first paradigm" (Howe 2014).

References

Akrich, M. (1987) "The De-Scription of Technical Objects," T. Pinch and W. E. Bijker (eds.) *The Social Construction of Technological Systems: New Directions in the Sociology and History of Technology*, Cambridge, MA: MIT Press, pp. 205–224.
Aradau, C. (2010) "The Myth of Preparedness," *Radical Philosophy* 161, pp. 2–7.
Bijker, W. E., R. Bal and R. Hendriks (2009) *The Paradox of Scientific Authority*, Cambridge, MA: MIT Press.
Bligh, W. G. (1910) "Dams, Barrages and Weirs on Porous Foundations," *Engineering News* p. 708.
Centraal, B. v. d. S. (2008) *Milieurekeningen 2008*, Den Haag and Heerlen, Netherlands: Centraal Bureau voor de Statistiek.
Daston, L. and P. Galison (2007) *Objectivity*, Cambridge, MA: MIT Press.
Disco, C. and L. van den Ende (2003) "'Strong, Invincible Arguments?' Tidal Models as Management Instruments in Twentieth-Century Dutch Coastal Engineering," *Technology and Culture* 44(3), pp. 502–535.
Gross, M. (2010) *Ignorance and Surprise*, Cambridge, MA: MIT Press.
Harteveld, C. (2011) *Triadic Game Design: Balancing Reality, Meaning and Play*, London: Springer.

Hilgartner, S. (2000) *Science on Stage: Expert Advice as Public Drama*, Stanford, CA: Stanford University Press.

Howe, J. (2014) *Behind the Curve. Science and the Politics of Global Warming*, Seattle, WA: University of Washington Press.

Kouw, M. (2015) "Standing on the Shoulders of Giants—and Then Looking the Other Way? Epistemic Opacity, Immersion, and Modeling in Hydraulic Engineering," *Perspectives on Science* 24(2), pp. 206–227.

Kouw, M., A. Scharnhorst and C. van den Heuvel (2013) "Exploring Uncertainty. Classifications, Simulations and Models of the World," P. Wouters, A. Beaulieu, A. Scharnhorst et al. (eds.) *Virtual Knowledge: Experimenting in the Humanities and the Social Sciences*, Cambridge, MA: MIT Press, pp. 89–125.

Latour, B. (1987) *Science in Action: How to Follow Scientists and Engineers through Society*, Cambridge, MA: Harvard University Press.

Latour, B. (1999) *Pandora's Hope: Essays on the Reality of Science Studies*, Cambridge, MA: Harvard University Press.

MacKenzie, D. A. (1999) "Nuclear Missile Testing and the Social Construction of Accuracy," M. Biagioli (ed.) *The Science Studies Reader*, London: Routledge, pp. 342–357.

Morgan, M. and M. Morrison (1999) "Models as Mediating Instruments," M. Morgan and M. Morrison (eds.) *Models as Mediators: Perspectives on Natural and Social Sciences*, Cambridge, MA: Cambridge University Press, pp. 10–37.

Nowotny, H. (2003) "Democratising Expertise and Socially Robust Knowledge," *Science and Public Policy* 30(3), pp. 151–156.

Oreskes, N., K. Shrader-Frechette and K. Belitz (1994) "Verification, Validation, and Confirmation of Numerical Models in the Earth Sciences," *Science* 263(5147), pp. 641–646.

Petersen, A. C. (2012) *Simulating Nature: A Philosophical Study of Computer-Simulation Uncertainties and Their Role in Climate Science and Policy Advice*, Boca Raton, FL: CRC Press.

Rijkswaterstaat (2005) *Veiligheid Nederland in Kaart: Hoofdrapport onderzoek overstromingsrisico's*, Den Haag, Netherlands: Ministerie van Verkeer en Waterstaat.

Vrijling, J. K., M. Kok, E. O. F. Calle et al. (2010) *Piping: Realiteit of Rekenfout? Expertise Netwerk Water*, (Rijkswaterstaat, Waterdienst), available from http://repository.tudelft.nl/islandora/object/uuid:f5b79879-f4d2-4fce-9b11-38547db4509f?collection=research, accessed August 2016.

Vrijling, J. K., A. C. W. M. Vrouwenvelder, M. Kok et al. (2004) *Review van de Resultaten van de Koplopers van 'Veiligheid van Nederland in Kaart'*, 3 May 2004.

Winsberg, E. (2006) "Models of Success Versus the Success of Models: Reliability Without Truth," *Synthese* 152, pp. 1–19.

Wynne, B. (1992) "Uncertainty and Environmental Learning," *Global Environmental Change* 2, pp. 111–127.

12 Tracing uncertainty management through four IPCC Assessment Reports and beyond

Catharina Landström

Introduction

Intriguing remarks

In conversations[1] with climate scientists from different areas about how they conceived of uncertainty in their work, many mentioned that experts specializing on the issue provided assistance. The climate scientists explained that to undertake uncertainty analyses sophisticated enough to get published in prestigious scientific journals, they relied on the knowledge and skill of "uncertainty experts." These were casual remarks, simply recognizing individuals who contributed to research projects. However, the remarks about experts on uncertainty prompted my curiosity and led to the exploration presented in this essay.[2]

Contemplating how to approach the idea of uncertainty experts and climate science, I decided to begin by looking at how uncertainty was described in the four Intergovernmental Panel on Climate Change (IPCC) Assessment Reports (AR) published at the time. Set up to summarize and assess climate change science for the world's governments, the IPCC has had a major impact on the evolution of climate science, including the way uncertainty is understood, addressed, and domesticated.

Uncertainty is much discussed within and outside of climate science. Written from the perspective of science and technology studies (STS), this essay adds to the discussion about how climate modeling, as represented in IPCC documents, has contributed to the evolution of "uncertainty management" (Shackley and Wynne, 1996) as an expert practice. The analysis begins by tracing uncertainty management through the first four IPCC ARs. From there it follows web links between experts, research projects, organizations, and publications to outline a loose network that pivots on the quantitative study of uncertainty. Finally, I summarize the findings and suggest a few lines of potential STS investigation arising from the study.

A science studies point of departure

From the perspective of the STS specialism of social studies of science, Simon Shackley and Brian Wynne's study of the rhetorical construction of uncertainty in the IPCC's communication of climate science to policy makers provides an important reference point. Shackley and Wynne devised a typology to explain the various ways in which uncertainty was narrated in the 1990 AR and the 1992 supplementary report. They distinguished four representations of "not knowing" by the IPCC, which they labeled "risk," meaning that the probability of a known event occurring is not known; "indeterminacy," which is present when not all parameters of a system and their interactions are fully known; "ignorance," pertaining to situations in which it is not known what is not known, and finally, "uncertainty," when enough is known to make qualitative but not quantitative judgments (Shackley and Wynne, 1996). They found that the IPCC emphasized "uncertainty" that was tractable and open to further research that could bring it into the realm of the quantifiable. The social scientists also identified rhetorical strategies, which transformed, and/or condensed, indeterminacy and ignorance into uncertainty.[3] Finding that uncertainty was articulated in several different ways in the IPCC documents, not only as error or lack of knowledge that was self-evidently reducible through more research, Shackley and Wynne (1996) introduced the notion of "uncertainty management" to capture the ongoing activity of controlling uncertainty in the communication of scientific knowledge. Shackley and Wynne's work laid the foundation for further social studies of science exploration of scientific uncertainty, and in the present essay their notion of uncertainty management as a distinct social activity provides the starting point for the following reading of the first four IPCC reports as traces of a continuous process in which this practice evolves.

FAR and SAR revisited

The First Assessment Report (FAR) of the IPCC Working Group (WG) 1, dedicated Chapter 11 to discussing uncertainty (Houghton et al. 1990). This chapter, entitled "Narrowing the Uncertainties: A Scientific Action Plan for Improved Prediction of Global Climate Change," describes climate science as a new research field, which is expected to improve, particularly with regard to "model skill," a notion borrowed from weather forecasting and referring to the ability of a model to give reliable forecasts. WG1 explains that climate predictions need to become more reliable and precise for society to prepare for the impact of climate change and it identifies the most important areas of uncertainty:

1 The sensitivity of global average temperature and mean sea levels to increases in greenhouse gases.
2 Regional climate impacts.
3 The timing of expected climate change.
4 Natural variation.

Uncertainties are defined here as remaining as problems or obstacles to be overcome. The chapter also talks about uncertainty resulting from a scarcity of global data, which motivates calls for more research. Other key issues are developing climate models with higher resolution and better ability to represent the ocean, as well as coupling the atmosphere with the ocean in models. The reader is warned not to expect visible progress until 1995 and after; what is more, global models incorporating chemical and biosphere aspects are not expected to materialize until the end of the 1990s.

The IPCC's Second Assessment Report (SAR) again has a WG1 chapter addressing uncertainty in very much the same terms as FAR (Houghton et al. 1996). The follow up of the uncertainties identified as reducible five years earlier, consists of two remarks: first, a declaration that major progress has been made in climate modeling with the inclusion of radiative forcing by aerosols, and second, that progress in climate science is to be expected on a five to ten-year scale and that it is important to keep up data collection and research in order for progress to become visible.

After these comments WG1 moves on to explain the scientific research process to the reader. A key assertion in this presentation is that more research is needed to advance the understanding of the climate system, and the United Nations (UN) Framework Convention on Climate Change is identified as the contextual driver for better scientific knowledge. Elaborating on what ought to be done, WG1 emphasizes the importance of reducing uncertainty in each individual discipline contributing to the understanding of climate. The authors also explain that the existence of uncertainties necessitates the presentation of ranges of likely values for climate variables, rather than definite singular numbers. An eventual reduction of uncertainties, defined as an ability to move the high and low values in a range closer together, is foreshadowed. A "framework" for the analysis of uncertainties is suggested, which lists six unknowns:

1   future greenhouse gas emissions;
2   future atmospheric concentrations of greenhouse gases;
3   results of additional radiative forcing;
4   the responses of the climate system to changed inputs;
5   how to distinguish between natural and human-induced changes in the climate; and
6   what the impact will be on sea levels and ecosystems.

The main part of the discussions of uncertainty in FAR and SAR pertain to a lack of knowledge due to scarcity of information of a kind that is already well understood, that is, gaps in data sets, which can be reduced by doing more empirical research. There is also some discussion of uncertainty that arises from future circumstances being unknown, such as greenhouse gas emission levels, which involve political decision-making. Shackley and Wynne's interpretation of this discussion as management of uncertainty, which formulates it in a way that makes it surmountable with more research and the unfolding of future social choices, was certainly comprehensive.

Data quality and quantity remain important sources of uncertainty, which climate scientists address in various ways. In conversation, an atmospheric chemist emphasized the importance of researchers' practical skills with using instruments (pers. comm. with an atmospheric chemist, 14 February 2011).This remark draws attention to the material aspects of climate science. Measurements must be reliable, and more data of better quality will result in some uncertainties being reduced. However, there are some uncertainties that cannot be reduced, even in principle, by improving measurements.The same atmospheric chemist explained that there would always be "atmospheric noise," that is substances or conditions affecting the instruments, such as dust from storms or variations in temperature, which make it critical to treat measurement results statistically to ensure that the "signal" is captured and the impact of "noise" is clearly understood. A similar view was expressed by a scientist specialized in satellite monitoring (pers. comm. with a satellite data expert, 3 February 2011). He talked about how the remote sensing data provided by satellites incorporate modeling and mathematical treatments from the outset, which makes the notion of data considerably more complex that is indicated in the discussions in FAR and SAR.

With the benefit of hindsight, I understand the discussions of uncertainty in FAR and SAR as focusing on uncertainties which continue to be regarded as reducible through more research, better instruments and improved techniques. However, the considerations in the third AR indicate a shift.

TAR and AR4 extending uncertainty management

In preparation for the Third Assessment Report (TAR), the IPCC commissioned two climate scientists, Richard Moss and Stephen Schneider, to write a guidance paper on uncertainty (Moss and Schneider, 1999). In the paper, Moss and Schneider advise the TAR contributors to adopt a "common approach for assessing, characterizing, and reporting uncertainties in a more consistent—and to the extent possible, quantitative—fashion across the various chapters" (1999: 35). Moss and Schneider's paper may have had a limited impact on TAR, as critics point to the lack of systematic discussion of uncertainty in it, but it is a sign that uncertainty management had attracted the attention of the IPCC.

Looking at WG1's contribution to TAR (Houghton et al., 2001) most of the uncertainties discussed are the same as in FAR and SAR, and considered to be reducible through more research. However, in TAR, WG1 also argues for the need to:

> *Improve methods to quantify uncertainties of climate projections and scenarios, including development and exploration of long-term ensemble simulations using complex models.*

> (Houghton et al. 2001: 771, emphasis in original)

Asking for the improvement of methods to quantify uncertainty in projections and scenarios is different from wanting to reduce uncertainty through more and better data. Quantification involves numerical estimation of the range of uncertainty associated with a projection or scenario. This is also a shift in attention from the uncertainty of the input (empirical data) provided to models, to the uncertainty of the output of climate models (projections and scenarios). This shift connects with WG1's discussions of new ways of working with climate models.

In TAR WG1 discusses a way of managing uncertainty not mentioned in FAR or SAR: comparison of the performance of different models and their components. In Chapter 8, entitled "Model Evaluation," the authors explain how coupled Atmosphere-Ocean General Circulation Models (AOGCM) had been evaluated against observations and how the differences between the models were interpreted. The chapter warns against assuming that credible simulation of present climate by a model guarantees that the "response to a perturbation remains credible" (Houghton et al. 2001: 473). WG1's assessment of model credibility is described in detail, with a table listing and comparing the AOGCMs, accompanied by numerous graphs and tables to illustrate the comparison process. In the final section the authors state that their "attempts to evaluate coupled models have been limited by the lack of a more comprehensive and systematic approach to the collection and analysis of model output from well co-ordinated and well designed experiments" (ibid.: 511). Despite this limitation, they proceed to declare levels of confidence in model results, which shows that the practice of model evaluation can facilitate a move from assessing uncertainty to declaring confidence. This way of defining uncertainty, quantitatively, to arrive at a claim about a degree of confidence, is a different type of uncertainty management than that discerned by Shackley and Wynne in the previous ARs.

The analysis of model ensembles in TAR focuses on quantifying discrepancies between model results and observational data. WG1 explains that climate models cannot be subject to falsification as defined in the philosophy of science, but they argue that it makes sense to try to establish whether a specific "model is 'unusable' in answering specific questions" (ibid.: 474). To accomplish this, the methods used in weather forecasting to evaluate model skill were considered useful. Further along in the text, the authors explain that there are, however, limits to what climate modelers can learn from weather forecasting since the latter "has the near-time reality of the evolving weather as a constant source of performance metrics" (ibid.: 775). In this discussion WG1 introduces a mathematical tool called a "Taylor diagram." This tool is said to enable analysis of how well models simulate the current climate, but it is not explained in any depth. The authors further argue that climate models "now have some skill in simulating changes in climate since 1850" (ibid.: 493). They also state that the simulation of past climates can be considered robust—all known changes are accounted for. However, WG1 does insist that regardless of a model's ability to simulate the past and the present, the accuracy in predicting the future cannot be established in the same way. This impossibility of knowing what will occur in the future provides a backdrop for the summarizing statement that "[A]n overriding challenge to modeling and to the IPCC is prediction" (ibid.: 775). Still, that the future is beyond the knowable does not mean that models cannot be improved; in the view of WG1: "[E]valuating the prognostic skill of a model and understanding the characteristics of this skill are clearly important objectives" (ibid.: 775). Managing uncertainty has taken a different direction in TAR, as it is much more about models than was the case in the previous ARs.

In the Fourth Assessment Report (AR4), published in 1997, ensemble modeling is discussed in more confident terms than in TAR. The Summary for Policymakers by WG1 states that:

> A major advance of this assessment of climate change projections compared with the TAR is the large number of simulations available from a broad range of models. Taken together with additional information from observations, these provide a quantitative basis for estimating likelihoods for many aspects of future climate change.
>
> (IPCC 2007: 12)

This statement summarizes a discussion about the evaluation of climate models in a coordinated comparison of the performance of different models. The details of this comparison are presented in Chapter 8, "Climate Models and Their Evaluation," discussing how "[E]ighteen modelling groups performed a set of coordinated, standard experiments, and the resulting model output" (Randall et al. 2007: 593).

At the time of the publication of AR4, climate scientists were continuing to write about model ensemble comparisons in scientific journals. In one paper, Mat Collins (2007) explains that there were three distinct types of comparisons, that is "ensembles":

1 multiple runs of one model with varying initial conditions;
2 multi-model ensembles comparing different models running the same experiment (also called "intercomparisons"); and
3 varying the parameters in the same model over many runs (i.e. "perturbed physics").

According to this typology the ensembles presented in TAR in 2001 were multiple runs, comparing how a model performed different simulations in relation to known past and present conditions. The ensembles discussed in AR4 in 2007 were intercomparisons, with each of the eighteen modeling teams performing the same experiments. The social dynamics of intercomparisons have been analyzed in social studies of science. Mikaela Sundberg (2011) investigated the social negotiations modelers undertook to establish what counted as a comparison. This did present a challenge when the objective was to compare and evaluate different models in which the non-linear, dynamic processes of physical systems are represented with different mathematical solutions and computer codes. The AOGCMs used to simulate climate comprise millions of lines of computer code, and it is not immediately obvious which of them make models different (or similar).

In AR4, perturbed physics was a less clearly defined type of ensemble; in one place, it was juxtaposed with intercomparisons as "ensemble model studies" (Le Treut et al. 2007: 118), and in another chapter (Meehl et al. 2007) it was discussed as one type of ensemble. In contrast Collins (2007) firmly defined a perturbed physics ensemble as running the same model with different parameters.

Perturbed physics is an approach expected to generate all simulation outcomes it is possible to achieve with one model (Frame et al. 2007). This method places great demands on computer capacity, it requires more memory and processing ability than what most university departments provide (Parker 2010). This computational constraint has prompted innovation that moves the computation outside of the scientific community to the public. Climateprediction. net, first introduced in an article in *Nature*, developed an approach in climate modeling similar to the projects enrolling home computers to analyze huge amounts of data from radio telescopes looking for signals possibly indicating the existence of intelligent life elsewhere in the universe (Allen 1999). The climateprediction.net project came on line in 2003 and it had become important enough to merit a brief discussion in WG1's Chapter 10 of AR4 (Meehl et al. 2007: 805, 806).

In TAR and AR4 model ensembles can be seen to have become a critical part of uncertainty management in the climate change science assessed by the IPCC. We can also see how addressing the uncertainty associated with the outcomes of simulations involves systematic analysis of large amounts of model-generated data, which requires rigorous mathematical treatments.

Bringing rigor to uncertainty management with frequentist and Bayesian production of PDFs

The large amounts of data generated in ensemble experiments, plus a growing body of data from observation of different climate variables, made it both necessary and interesting to develop statistical methods to trace patterns in model behavior. Appendix A to Chapter 9 in AR4 (Hegerl et al. 2007) discusses the statistical methods used to estimate the probabilities of the simulation outcomes for different climate variables. The authors begin by stating that the IPCC mainly relied on what they call "frequentist" statistical methods.

"Frequentist" uncertainty analysis is often explained by making a simile with tossing a coin. It is possible to statistically predict the relative frequency of heads and tails over a large number of tosses, although it is not possible to predict the exact outcome of the next toss. Statistical "Monte Carlo methods" use the possibility of predicting the frequency of particular outcomes over a large number of computer model runs to assess the reliability of projections.

The type of uncertainty that can be quantified through frequentist methods is also called "aleatory" and understood to reside in the phenomena investigated and not to be reducible by increasing knowledge about the possible outcomes (pers. comm. with a mathematician, 24 January 2011).

Another way of addressing uncertainty mentioned in AR4 is called "Bayesian," although the authors found it to be unusual at the time, explaining that:

> Bayesian approaches are of interest because they can be used to integrate information from multiple lines of evidence, and can incorporate independent prior information into the analysis. [...] inferences are based on a posterior distribution that blends evidence from the observations with the independent prior information, which may include information on the uncertainty of external forcing estimates, climate models and their responses to forcing. In this way, all information that enters into the analysis is declared explicitly.
>
> (Hegerl et al. 2007: 745)

Hegerl et al. (2007) explain that frequentist approaches can be difficult to apply to natural processes because they require that the probability of all possible outcomes be calculated for every event; this is not a problem in coin tossing with two options but when applied to the output data of multiple simulations of global climate it becomes very challenging. Another problem is that all outcomes may not be equally likely, for example, temperature trends could affect the likelihood of a succeeding event. In contrast to frequentist methods, Bayesian approaches do not assume that natural systems work as if every outcome is independent from the previous ones. Bayesian analyses allow for knowledge about priors to be taken into account (for an explanation of the Bayesian probability in an accessible way see Joyce 2003; Bonilla 2009).

Since AR4 was published, Bayesian uncertainty treatments have become more widely used in environmental modeling. Hydrologist Keith Beven describes the approach as "a form of statistical learning process" (2009: 152). In Bayesian analyses of model projections, more knowledge about climate processes, and better representation of these processes in models, will make a difference for the estimation of uncertainty. One climate scientist I talked to called the uncertainties captured with Bayesian approaches "epistemic," explaining that they originated in a lack of knowledge about the natural processes of climate, rather than the characteristics of these processes themselves (pers. comm. with a physicist, 28 February 2011).

Regardless of whether the approach is frequentist or Bayesian, statistical treatments of model uncertainties are used to produce probability distribution/density functions (PDFs). PDFs are tools for analyzing the behavior of random continuous variables; they enable calculation of the likelihood of a variable taking on certain values over others and the outcomes can be visualized in graphic plots. Moss and Schneider's guidance notes to TAR foreshadowed PDF visualization of quantitative assessments of uncertainty in talking about the possibility of drawing a "cumulative distribution function" (Moss and Schneider 1999: 41). At the time, they found that such analysis was possible in only a few cases, but five years later a report from a 2004 IPCC workshop on uncertainty (Manning et al. 2004) note that the use of PDFs is growing and explains that they "provide detailed quantitative descriptions of uncertainties" (ibid.: 6). In the guidance note for AR4 (IPCC 2005), prepared after the 2004 workshop, PDFs are mentioned as the final item on a list prescribing how to use the appropriate level of precision when describing research findings:

> *A probability distribution can be determined for changes in a continuous variable either objectively or through use of a formal quantitative survey of expert views*: Present the PDF graphically and/or provide the 5th and 95th percentiles of the distribution. Explain the methodology used to produce the PDF, any assumptions made, and estimate the role of structural uncertainties.
>
> (IPCC 2005: 3, original emphasis)

In AR4, PDFs are applied to projections of future global climate states in WG1's Chapter 9 (Meehl et al. 2007). Different ways to produce PDFs are described, ranging from the purely statistical to methods which make extensive use of observation data. In the chapter, PDFs figure as key tools, allowing for model outcomes to be analyzed and compared:

> Additionally, projection uncertainties increase close to linearly with temperature in most studies. The different methods show relatively good agreement in the shape and width of the PDFs, but with some offset due to different methodological choices. Only Stott et al. [...] account for variations in future natural forcing, and hence project a small probability of cooling over the next few decades not seen in the other PDFs. The results of Knutti et al. [...] show wider PDFs for the end of the century because they sample uniformly in climate sensitivity [...] Resampling uniformly in observables would bring their PDFs closer to the others.
>
> (Meehl et al. 2007: 809)

This quote demonstrates the adoption of PDFs by the IPCC authors as a way to compare and communicate the degree of uncertainty in climate model simulation outcomes. In the period between TAR and AR4, PDFs became a common statistical tool for describing uncertainties in climate model outputs. This was also how they were talked about; one scientist, who specialized on modeling the chemical processes of climate, regarded the presentation of the uncertainty of model outputs in PDFs as routine (pers. comm. with an atmospheric chemist, 3 February 2011). The use of Bayesian treatments to generate PDFs were also becoming commonplace. When writing in 2007, Collins regarded Bayes' theorem, articulated as a statistical probability equation, as a regular way to generate PDFs for climate models.

This reading of FAR, SAR, TAR and AR4 has traced uncertainty management as a process evolving over time, thereby bringing to the forefront shifts of emphasis in WG1's discussions. In the early reports uncertainty management was focused on reducing uncertainty through more empirical research to improve the quantity and quality of data, thus it was a feature motivating more empirical research on climate processes. In TAR and AR4 the focus was on uncertainties generated in the process of simulation modeling which were addressed with statistical tools, prompting quantitative rigor in the management of uncertainty.

The shift in focus to uncertainties arising from computation of climate processes turns around the problem presented by data, from data scarcity to data abundance. It is the very large amounts of output data generated by model simulations that demand sophisticated statistical methods for understanding the uncertainties they contain. The opportunity to apply sophisticated mathematical methods to analyze and represent uncertainty in climate models has attracted interest from scientists with expertise in statistics, rather than climate. In the next section I will trace a few of these uncertainty experts in cyberspace.

Model-based uncertainty analysis beyond the IPCC

From reading the IPCC ARs as traces of a process, I learned that uncertainty management has evolved into sophisticated quantitative analysis, and I knew from conversation with climate scientists that they rely on assistance by experts on uncertainty analysis. I was by now very curious about these experts. While it was beyond the scope of the study presented in this essay to undertake a proper study of uncertainty expertise, it was possible to do some tentative exploration on the web. The starting point was uncertainty experts named by the climate scientists in informal conversations.

One person mentioned by several climate scientists was Jonathan Rougier. Online it was possible to find out that he is a statistician with expertise in Bayesian uncertainty treatments, working at the University of Bristol in the Department of Mathematics. He describes his research interests as focusing on the uncertainty inherent in complex systems like the climate. He also provides a list of publications, beginning in 1993 with an article on margin traders on the London Stock Exchange. In 2001 he published a paper on "Bayesian forecasting for complex systems using computer simulation" and in 2004 a paper on server advantage in tennis; in 2007 he wrote about climate models, first in a paper on probabilistic inference for future climate using an ensemble of climate model evaluations. This list of publications demonstrates that Rougier's expertise cuts across very different empirical fields, and indicates the independence of uncertainty analysis from the uncertainty management undertaken in climate science. From the list of publications, it is also clear that in addition to contributing to uncertainty management in empirical research fields, Rougier collaborates with other researchers on developing the mathematical analysis of uncertainty.

One of these collaborators is David Stainforth who, according to his presentation on the London School of Economics and Politics (LSE) website, is a physicist with many years' experience in climate modeling, currently working at the Grantham Institute. Stainforth describes his interests as concerning how to extract robust and useful information about future climate from modeling and how to design modeling experiments that can be linked to real-world decision making. His publications from 2008 and onward cover academic articles focusing on uncertainty in climate science, such as "Estimating uncertainty in future climate projections" (2010), and discussions of uncertainty, which are less technical (mathematical) and more generally accessible, in non-specialist media, such as broadsheet newspapers. This shows that expertise on uncertainty can generate knowledge of interest to wider society.

One of Stainforth's mainstream media articles, "Policy: clarify the limits of climate models" (Stainforth and Smith 2012), is co-authored with Leonard A. Smith, another person recurrently referred to as an "uncertainty expert" in conversations with climate scientists. Smith's electronic trace leads to the Department of Statistics at LSE, his "experience keywords" include statistical analysis of large computer models, predictability and analysis of significance in statistical estimates. His list of publications reaching back to 1980 covers a wide range of issues with titles such as the co-authored papers "Improved analytic characterization of ultraviolet skylight" (Green et al. 1980), "A method for generating an artificial RR tachogram of a typical healthy human over 24-hours" (McSharry et al. 2002) and "On virtual observatories and modeled realities (or why discharge must be treated as a virtual variable)" (Beven et al. 2012); in addition, he has many publications on weather forecasting and climate modeling.

The simple online tracing of these three scientists identified as uncertainty experts showed that they had different disciplinary backgrounds and different research focuses: One specialized in Bayesian analysis; another in designing model experiments, and the third in statistically distinguishing signal from noise. The link between them is an interest in researching uncertainty using mathematical methods. Their lists of publications show that they collaborate with many other scientists on the quantitative examination of uncertainty and on refining their analytical tools.

From this very tentative online tracking of people who were mentioned in informal conversations with climate scientists based in the United Kingdom, it is not possible to draw any conclusions other than that it would be interesting to conduct a proper empirical study of uncertainty experts to develop a social studies of science understanding of their knowledge, skills, and activities. However, without making any claims about these three UK uncertainty experts, it is possible to take the web tracking further by following links from their websites.

Such web links lead to collaborative research efforts. One of these called "MUCM," was described as "a research project concerned with issues of uncertainty in simulation models (also known as process models, mechanistic models, computer models, etc.)" (MUCM 2015). The MUCM project was concluded at the end of 2012, after running for over six years. The project website talks about a range of techniques deployed in the investigations, for example, "uncertainty quantification," "propagation," and "analysis." It is also explained that after the conclusion of the project the researchers involved continue to collaborate in an online "MUCM Community," which maintains and uses the website and the archive (MUCM 2015a). The creation of a community of researchers was one stated objective of the project, another was to build technology "that bridged the gap from research to application and made the methods of quantifying model uncertainty available to the wide community of modelers and model users, as well as other researchers" (MUCM 2015b). This project appears to also have embodied the two aims defining expertise—to understand an object of investigation and to deploy the knowledge generated to assist others.

Many of the links on the MUCM website lead to similar research initiatives in different countries, but one connects to another discipline: computer science. The European Exascale Software Initiative (EESI) involves software researchers in universities and institutes across Europe and is co-funded by the European Commission. The objective of EESI 2, 2012–15 was to "build a European vision and roadmap to address the challenges of the new generation of massively parallel systems composed of millions of heterogeneous cores 'which will provide multi-Petaflop performances in the next few years and Exaflop performances in 2020" (EESI 2013). One of the many topics this project will address is "uncertainty," and although it is only a small part of this very large project, the link from MUCM to the EESI draws attention to the relationship between scientific uncertainty and the computers that make simulation modeling possible.

Another web link leads to a recently finished project in the United States that shows the involvement of a wide range of formal institutions. In 2011 and 2012 the Statistical and Applied Mathematical Sciences Institute (SAMSI), a collaboration involving Duke University, North Carolina State University, University of North Carolina at Chapel Hill, and the National Institute of the Statistical Sciences, ran a program on "Uncertainty Quantification" with a sub-program on climate modeling. SAMSI's mission is "to forge a synthesis of the statistical sciences and the applied mathematical sciences with disciplinary science to confront the very hardest and most important data- and model-driven scientific challenges" (SAMSI 2015). Presented as a response to the growth and spread of simulation modeling across many different fields the uncertainty quantification program focused on uncertainties arising from:

1   The gap between model and real system.
2   The cost of running models that constrain the availability of model outputs rather than a one-to-one correspondence with inputs.
3   The uncertainties of initial conditions, unknown parameters in models, stochastic model features.
4   The need to combine model output and noisy data for predictions.

Further links to scientists involved with the uncertainty quantification program lead to the Institute for Mathematics Applied to Geosciences (IMAGe), presented as a division of the Computational & Information Systems Laboratory (CISL), which brings mathematical models and conceptual tools to bear on fundamental problems in the geosciences, and said to be envisaged to grow as a center of activity for the mathematical and geophysical scientific communities. One IMAGe working group is called the Regional Integrated Science Collective (RISC) and aims to, among other things, analyze the uncertainty inherent in climate change on the regional scale. The current focus is the North American Regional Climate Change Assessment Program (NARCCAP), which produces multi-model ensembles of high-resolution future climate simulations to explore uncertainties related to the spatial scale of climate simulations.

While it is clear that some of the research projects focusing on the quantitative study of uncertainty have ended, this brief web mapping does indicate continuing institutional interest in the topic. The web links between people, activities, and research organizations do show that efforts have been and continue to be made to make connections among actors interested in examining scientific uncertainty associated with environmental modeling. Online connections do not tell us about offline relationships, whether the projects presented on web sites have led to the institutionalization of research on uncertainty, or whether courses on the topic have become embedded in degree programs. Further historical and social science investigation is necessary to form a view on this, but the web links show that there was substantial interest in researching uncertainty quantitatively, something also expressed in the existence of academic journals specializing in the area.

Web links also led to two recently launched academic journals. The *International Journal for Uncertainty Quantification* states its aim as disseminating "information of permanent interest in the areas of analysis, modeling, design and control of complex systems in the presence of uncertainty" (2016). The journal aims to cover different methods—"stochastic analysis, statistical modeling and scientific computing"—all used to address uncertainty. Its territory is broadly defined, covering climate models and other complex dynamic systems mathematically represented which are "governed by differential equations." The journal identifies the "representation of uncertainty, propagation of uncertainty across scales, resolving the curse of dimensionality, long-time integration for stochastic PDEs, data-driven approaches for constructing stochastic models, validation, verification and uncertainty quantification for predictive computational science, and visualization of uncertainty in high-dimensional spaces" as topics of particular interest. It also encourages submissions addressing "applications of uncertainty quantification in all areas of physical and biological sciences" (*International Journal for Uncertainty Quantification* 2016).

The second journal has a similar title *SIAM/ASA Journal on Uncertainty Quantification*. Started in 2013, the web presentation refers to an "increasing importance of uncertainty quantification (UQ) in science and society" (*SIAM/ASA* 2013). This increase is understood to be the "consequence of the central role of mathematical, statistical, and computational modeling in scientific discovery, engineering design, risk assessment, and decision making in almost every area of human activity" (*SIAM/ASA*, 2013). The first issue of this fully electronic journal was published in March 2013.

These two new journals would provide interesting starting points for more extensive social studies of science exploration of the quantitative study of scientific uncertainty, regardless of their continuation and success. Even if both journals should close after a few issues, their very existence demonstrates a quantitative approach to uncertainty that surpasses the understanding of uncertainty management currently dominant in the social studies of science.

Conclusions and suggestions for future research

The exploration presented in this essay was prompted by curiosity about remarks made in informal conversations with climate scientists about experts on uncertainty. Starting with the notion of uncertainty management introduced by Shackley and Wynne in a study of the first IPCC AR and supplementary documents, I set out to look at WG1's contributions to the first four IPCC reports as a continuous sequence. This examination showed that the quantitative management of uncertainty arising in the computer simulation modeling of climate, which played a minor part in the material studied by Shackley and Wynne, had, over time, moved to the center of the discussion and evolved into sophisticated statistical approaches. At the time of AR4 the widely established mathematical tool of PDFs were routinely used to analyze and visualize the uncertainties of model ensemble outcomes. It is at this time and later that experts on uncertainty treatments become visible to this exploration through mentions in conversations as contributors to managing uncertainty in modeling projects. By following traces from written documents and spoken comments to the Internet, I found virtual links between experts, research projects, institutions and journals connected by quantitative analysis of the uncertainty associated with computer simulation modeling. In this essay, I have tentatively suggested that the traces of actors and activities on the Internet can be interpreted as indicating that uncertainty may be emerging as a topic of study, independent of any particular model or modeling experiment.

The results of this essay show that the domestication of uncertainty in climate prediction was afforded increasing attention and increasingly professional treatment by the IPCC. The framing of uncertainty in the first two assessment reports as a problem to be overcome with more data and better models met with consensus among those scientists involved in the preparation of report chapters. This treatment, however, apparently failed to satisfy IPCC audiences sufficiently and in a sustained manner. Discourses about uncertainty as an apparent weakness of climate change science stayed with the IPCC and continued to pose threats to its scientific and public authority. Domesticating uncertainty demanded a more professional and consistent approach and more careful management of its rhetorical and graphic representation. The involvement of specialized uncertainty experts and sophisticated expertise inaccessible to outsiders represents an exemplary approach to domesticating uncertainty in such cultures of prediction as climate science. The public character of climate science, however, is likely to make it difficult to achieve a "black-boxing" of uncertainty issues as in other domains (see also Chapter 2 and Kouw in this volume).

This exploration was a modest beginning and needs to be continued to account for the further development of IPCC treatments of uncertainty in AR5, as well as in the contributions of WG2 and WG3, the discussions of uncertainty in which may lead in different directions from that of WG1. The informal conversations with climate scientists featured in this essay warrant expansion into a comprehensive interview study with a larger number of climate scientists, uncertainty experts, and other relevant constituents. Finally, the connections between people, projects, and institutions found in cyberspace could be interrogated in relation to bibliometric information to thoroughly map networks of uncertainty expertise.

Notes

1　These conversations were informal background briefings used in this essay to indicate ideas that some climate scientists held at the time (2011). It is important to distinguish such conversations from formal social science interviews undertaken with the aim of systematic analysis.

2　Defining this text as an essay sets it apart as using a more narrative form than a research article. The essay format allows for the successive introduction of elements as the exploration moves forward, while a research article usually requires comprehensive presentation of the material, methods and questions at the outset.

3　Stirling and Gee (2002) have refined and developed this typology further, paying particular attention to the ways in which scientists have tried to present uncertainty of all types as quantifiable risk in communication with policy makers.

References

Allen, M. (1999) "Do it yourself climate prediction," *Nature* 401, p. 642.

Beven, K. (2009) *Environmental Modelling: An Uncertain Future*, London: Routledge.

Beven, K., W. Buytaert, and L. A. Smith (2012) "On virtual observatories and modeled realities (or why discharge must be treated as a virtual variable)," *Hydrological Processes* 26(12), pp. 1905–1908.

Bonilla, O. (2009) *Visualizing Bayes Theorem*, available from https://oscarbonilla.com/2009/05/visualizing-bayes-theorem/, accessed 20 February 2017.

Collins, M. (2007) "Ensembles and probabilities: A new era in the prediction of climate change," *Philosophical Transactions of the Royal Society A* 365, pp. 1957–1970.

Edwards, P. N. (2010) *A Vast Machine. Computer Models, Climate Data, and the Politics of Global Warming*, Cambridge, MA: MIT Press.

Edwards, P. N. and S. H. Schneider (2001) "Self-governance and peer review in science-for-policy: The case of the IPCC second assessment report," C. A. Miller and P. N. Edwards (eds.) *Changing the Atmosphere. Expert Knowledge and Environmental Governance*, Cambridge, MA: MIT Press.

EESI European Exascale Software Initiative (2013) EESI2 presentation on CINECA website at http://hpc.cineca.it/projects/eesi2, accessed 20 February 2017.

Frame, D. J., N. E. Faull, M. M. and Joshi (2007) "Probabilistic climate forecasts and inductive problems," *Philosophical Transaction of the Royal Society A* 365, pp. 1971–1992.

Green, A. E. S., K. R. Cross and L. A. Smith (1980) "Improved analytic characterization of ultraviolet skylight," *Photochemistry and Photobiology* 31(1), pp. 59–65.

Hegerl, G. C., F. W. Zwiers, P. Braconnot et al. (2007) "Understanding and attributing climate change," S. Solomon, D. Qin, M. Manning et al. (eds.) *Climate Change 2007: The Physical Science Basis*, Cambridge, MA: Cambridge University Press.

Houghton, J. T., G. J. Jenkins, and J. J. Ephraums (eds. 1990) *Climate Change 1990: The IPCC Scientific Assessment. Contribution of Working Group I to the First Assessment Report of the Intergovernmental Panel on Climate Change*, Cambridge, MA: Cambridge University Press.

Houghton, J. T., L. G. Meira Filho, B. A. Callander et al. (eds. 1996) *Climate Change 1995: The Science of Climate Change. Contribution of Working Group I to the Second Assessment Report of the Intergovernmental Panel on Climate Change*, Cambridge, MA: Cambridge University Press.

Houghton, J. T., Y. Ding, D. J. Griggs, et al. (eds. 2001) *Climate Change 2001: The Scientific Basis. Contribution of Working Group I to the Third Assessment Report of the Intergovernmental Panel on Climate Change*, Cambridge, MA: Cambridge University Press.

International Journal for Uncertainty Quantification (2016) *Homepage*, available from http://www.begellhouse.com/journals/uncertainty-quantification.html, accessed April 2016.

IPCC (2005) *Guidance Notes for Lead Authors of the IPCC Fourth Assessment Report to Addressing Uncertainties*, Geneva, Switzerland: IPCC.

IPCC (2007) "Summary for policymakers," S. Solomon, D. Qin, M. Manning, Z. Chen, M. Marquis, K.B. Averyt, M. Tignor and H.L. Miller (eds.) *Climate Change 2007: The Physical Science Basis*, Cambridge, MA: Cambridge University Press.

Joyce, J. (2003) "Bayes' Theorem," *Stanford Encyclopedia of Philosophy*, available from http://plato.stanford.edu/entries/bayes-theorem/, accessed August 2016.

Le Treut, H., R. Somerville, U. Cubasch, Z. Chen, M. Marquis, K.B. Averyt, M. Tignor and H.L. Miller (2007) "Historical overview of climate change," S. Solomon, D. Qin, M. Manning, Z. Chen, M. Marquis, K.B. Averyt, M. Tignor and H.L. Miller (eds.) *Climate Change 2007: The Physical Science Basis*, Cambridge, MA: Cambridge University Press.

McSharry, P. E., G Clifford, L. Tarassenko, and L. A. Smith (2002) "A method for generating an artificial RR tachogram of a typical healthy human over 24-hours," *Computers in Cardiology* 29, pp. 225–228.

Manning, M., M. Petit, D. Easterling, J. Murphy, A. Patwardhan, H. -H. Rogner, R. Swart and G. Yohe (2004) *Workshop Report: IPCC Workshop on Describing Scientific Uncertainties in Climate Change to Support Analysis of Risk and of Options*, Geneva, Switzerland: IPCC.

Meehl, G. A., T. F. Stocker, W. D. Collins, et al. (2007) "Global climate projections," S. Solomon, D. Qin, M. Manning Z. Chen, M. Marquis, K.B. Averyt, M. Tignor and H.L. Miller (eds.) *Climate Change 2007: The Physical Science Basis*, Cambridge, MA: Cambridge University Press.

Moss, R. H. and S. H. Schneider (1999) *Uncertainties in the IPCC TAR: Recommendations to Lead Authors For More Consistent Assessment and Reporting*, Geneva, Switzerland: IPCC.

MUCM Managing Uncertainty in Complex Models (2015) *Homepage*, available from http://www.mucm.ac.uk/indexMUCM.html, accessed April 2015.

MUCM Managing Uncertainty in Complex Models (2015a) *Homepage*, available from http://www.mucm.ac.uk, accessed 22 April 2015.

MUCM Managing Uncertainty in Complex Models (2015b) *Homepage Community*, available from http://www.mucm.ac.uk/Pages/MCSGCommunity.html, accessed April 2015.

Parker, W. S. (2010) "Predicting weather and climate: Uncertainty, ensembles and probability," *Studies in History and Philosophy of Modern Physics* 41, pp. 263–272.

Randall, D. A., R. A. Wood, S. Bony, et al. (2007) "Climate Models and Their Evaluation," S. Solomon, D. Qin, M. Manning et al. (eds.) *Climate Change 2007: The Physical Science Basis*, Cambridge, MA: Cambridge University Press.

SAMSI Statistical and Applied Mathematical Sciences Institute (2015) *Homepage*, available from http://www.samsi.info/about/what-samsi, accessed April 2015.

Shackley S. and B. Wynne (1996) "Representing Uncertainty in Global Climate Change Science and Policy: Boundary-Ordering Devices and Authority," *Science, Technology & Human Values* 21(3), pp. 275–302.

SIAM/ASA Society for Industrial and Applied Mathematics (2013) *Homepage*, available from http://epubs.siam.org/doi/abs/10.1137/130973363, accessed April 2015.

Solomon, S., D. Qin, M. Manning, et al. (2007) "Technical summary," S. Solomon, D. Qin, M. Manning, Z. Chen, M. Marquis, K.B. Averyt, M. Tignor and H.L. Miller (eds.) *Climate Change 2007: The Physical Science Basis*, Cambridge, MA: Cambridge University Press.

Stainforth, D. A. (2010) "Estimating uncertainty in future climate projections," J. Rolf Olsen, J. Kiang and R. Waskom (eds.) *Workshop on Nonstationarity, Hydrologic Frequency Analysis, and Water Management*, (Colorado Water Institute Information 109), Fort Collins, CO: Colorado State University.

Stainforth, D. A. and L. Smith (2012) "Policy: clarify the limits of climate models," *Nature* 489, p. 208.

Stirling, A. and D. Gee (2002) "Science, precaution and practice," *Public Health Reports* 117, pp. 521–533.

Sundberg, M. (2011) "The dynamics of coordinated comparisons: How simulationists in astrophysics, oceanography and meteorology create standards for results," *Social Studies of Science* 41(1), pp. 107–125.

13 The future face of the Earth

The visual semantics of the future in the climate change imagery of the IPCC

Birgit Schneider

Visions of the future

In 1979, a two-page comic strip titled "A Short History of America" was published in an American magazine known as *CoEvolutionary Quarterly*.[1] In a space of only twelve panels, and without any text or commentary, cartoonist Robert Crumb condensed North America's development from the arrival of the first Europeans to his present day in the 1970s. Crumb's pictorial narration begins with an image of an untouched forest meadow successively transformed in the subsequent panels into an infrastructure dominated by smoking trains, highways, trams, antique cars and, ultimately, the Cadillacs of the 1970s. In the ninth panel, the last tree standing makes way for new streetlamps, traffic lights, telephone wires and advertising signs. Two years after his comic was first published, Crumb coloured it in and added three more panels (see Figure 13.1), each of which illustrates a prototypical vision of the future in an especially vivid manner. The first version has the title "Worst-case scenario, ecological disaster" and most clearly picks up on the previous panels. The street crossing visible in the initial panels is now depicted after an undefined ecological catastrophe. The history of modern lifestyle has come to an end: the sky is yellow, civilization is in ruins and it would appear that human existence has been wiped out. The second vision is called "The fun future: Techno-Fix on the march." Here, colourful cars fly across the blue sky and over curve-lined futuristic homes in clean, manicured gardens where "homo faber" appears to have everything under control. The third version shows "The ecotopian solution." The inhabitants of this future world shun the use of all electric and fuel-driven machines. In fact, they have returned to the woods, the lush sprawl of which recalls the forests depicted prior to the invasion of the Europeans. Here, the principles of evolution and growth have returned to a colourful prehistoric commune.

Interestingly, the three future scenarios posited in such a concise manner by Crumb also serve to convey the three ideal-typical visions of the future that arise in the history of modernism and its scientific and technical progress. Indeed, this history has produced options for the future that involve either catastrophic collapse, technical optimism—that is, homo faber's tireless

Figure 13.1 Three possible futures. "A Short History of America", 1 Worst-case
          scenario, ecological disaster, 2 The fun future: Techno-Fix on the march,
          3 The ecotopian solution.

Source: Comic by Robert Crumb for *Coevolution Quarterly*, 1979, 1981, coloured by Peter
Poplaski and distributed as a poster by Kitchen Sink Press. © Agence Litteraire Lora Fountain &
Associates.

discovery of technical solutions—or the principle of shrinkage, which involves
doing without and returning to a natural state. All three visions appear plausible and have their own group of followers depending on whom one asks and
whether one sees "nature" as being well-intentioned, unpredictable, delicate
or robust (Douglas and Wildavsky 1982; Hulme 2009: 186). In other words,
today, almost forty years after the comic was published, our current visions
of the future of a world in the grips of climate change still unfold within the
typology of these three apparitional visions.

Futuristic image types

The futuristic image type—one that illustrates possible futures in the form of a triptych—was originally developed for the purpose of depicting religious devotional images. For example, in the case of a winged altar from the fifteenth century that bears the title "The Last Judgment," we see, on the right panel, a highly dramatic portrayal of the descent of the damned into hell.[2] This large-scale altar piece was created by Hans Memling for a chapel in Italy as part of a commission he received from the Medici Family. In the triptych, the people who have been chosen to enter paradise are found on the left-hand panel and make their way naked up the stairs to the magnificent heavenly gate; the damned, on the other hand, are found on the right-hand panel and are being pushed by a number of devils into the glowing and flaming depths of hell. Their falling bodies form a chaos of limbs and faces deformed by pain and horror. It is the red shades of purgatory that dominate the right section of the image—they symbolise the eternally destructive force of fire.[3]

In our present time, images of a destructive and menacing future are provided by the imagined worlds created by climate statistics and cartography. They are drawn in the contemporary language of diagrams and maps and belong to a new "culture of prediction" in which the major issues facing the earth, such as ecological boundaries and climate change, are found. For this reason, it is necessary to take these images seriously—no matter how visually sparse they might be—and examine them precisely. Indeed, they inspire a highly important question: "How do visions of the future intervene in reality and what forms do they offer (and have offered historically) that allow us to perceive and structure reality?" (Horn 2014: 24).

They speak to the question, raised in Chapter 2, of the broader cultural impact of the environmental sciences' new cultures of prediction, in terms of influencing more widely distributed means of comprehending collective futures. Although the graphic curves and charts produced by climate research depict an entirely differently form of acquired knowledge than that portrayed by religious images, this essay seeks to argue that the colour semantics generated by science contribute in a fascinating and complex way—specifically by means of the fiery red colourings of an overheated planet—to images of final judgment, catastrophe, apocalypse and the end of days. The collective imaginary sees "burning worlds" in those red charts and curves and therefore evidence of an end to the world—an end, however, that is no longer depicted in a healing historical context.[4] In addition, science-based depictions of the future also continue to contain traditional approaches to visions of the future that are deeply anchored in the collective imaginary. At the same time, these image types serve to perpetuate the arrangement of images as diptychs or triptychs to be able to compare possible futures in the rational abstraction of coloured charts and curves.

Building on that thought, the key question contained in this essay is the following: How does the Intergovernmental Panel on Climate Change (IPCC) draw an image of possible futures in its reports? In addition to data presented in table form, there are three image types contained in the summarizing reports of the IPCC: curves, charts and diagrams. In its Summaries for Policy Makers (SPM), the IPCC summarises its key findings and arguments in the semiology of these charts and diagrams. In each of the summarised reports issued by the three Working Groups (WG I-III), there are roughly ten charts that can represent the depiction of the most relevant knowledge about climate change and its risks available today. The importance of images of the future in this context is demonstrated by the number of charts that depict time periods in the future. Over one third of the charts contained in the SPMs in the 2013/2014 report thematise possible futures.[5] In other words, these presentations aimed at policymakers reflect the extent to which climate research has transformed into a producer of future knowledge, that is, the extent to which it now engages in futurology. This future knowledge is created in the modern neo-liberal "regime of anticipation", which is based on the notion of a "management of time" (Adams et al 2009).

To be able to pursue questions relating to the aesthetics and structure of these images, I will use image-historical and art-historical methods to critically examine precisely what kind of image of the future is scientifically constructed in the charts published by the IPCC. In the course of this examination, relationships to other ideas and images of the future will also be established. My analysis of the images is based on three initial observations, all of which point to the instrumental character of images:

1   The future itself cannot be seen. Therefore, images that put forward possible futures represent an important way of being able to render the future imaginable in the first place—even if these images can only act as a base for the imagination.
2   As visions of the future, images play a central role, because they make it possible to have a discussion about desirable futures—but also about acceptable risks and fears—on a social level. In this sense, they function as "decision-making tools" (Rosentrater 2010), but also as a way to test our visions of possible futures.
3   They are also important in a general sense, seeing as images hold the potential power to change reality itself by means of people's imagination, that is, by changing how people think and make decisions about the future.

When, in the first section of this paper in particular, the analysis of the images takes into account visions of the "future as catastrophe" (Horn 2014; Walter 2010), this is due to the fact that these images can be linked so successfully to countless diagrams and charts on climate change. In this process, the analysis separates the charts from their captions and framing texts, because this is precisely what happens in the common practice of image dissemination. In other words, whether the authors desire so or not, images are often separated from all of the information contained outside of the image frame and, in effect, begin to speak "for themselves" and to be distributed independently from the framing texts.[6]

With our backs to the future

To what extent can we even imagine the future? How far into the future—in terms of time and geography—can we see? And how can we apply this to our knowledge of climate change? In her iconographic analysis, Kate Manzo divided up images of climate change into the categories of "fingerprints" and "harbingers", which can be most clearly applied to photographs (Manzo 2010: 97). Photographs such as those of melting glaciers represent fingerprints, while harbingers are represented by documentations of extreme weather events and their consequences, such as storms and floods. Images such as these reveal the already existing consequences of climate change and use present indicators to depict an outlook on the changing patterns and dimensions of future weather events.

The scenarios for the future that are produced by climate sciences belong to a third category. In a manner that differs from photography, they have no indexical relationship to that which is depicted as a fingerprint of actual reality. A visualization of the future cannot be undertaken by documentary means; instead, fictional means grounded in probability theory must be used. Seeing as direct access to the future is impossible, knowledge of the future is created by means of numeric simulations. Indeed, although the resulting curves and charts point like harbingers to future climate change (they even contain sections of the present day for the purpose of "initialising" the models), the media and probabilistic methods involved in this knowledge are simulation, scenario development and projection.[7]

The *Limits to Growth* report issued by the Club of Rome in 1972 can be seen as one of the first attempts to calculate possible futures in the subsequent decades using a world model as a computer simulation. Club of Rome researchers modelled the changing relationships between human activities and ecosystems using the World3 programme developed at the Massachusetts Institute of Technology. For their system, they used the following five variables: population growth, industry, environmental pollution, food production and resource consumption.[8] Based on these "main protagonists" in earth-related events, the research group created for the very first time a global ecological narrative of a possible future, including human factors, in the form of a model based on numbers and formulas. The book was highly circulated and can therefore be considered one of the first publications to sketch future narratives within the framework of statistical calculations and to communicate these in the form of a science-based warning to the general public.

At the beginning of the book, which otherwise contains several charts and feedback diagrams, is a chart with the title "Human Perspectives" (see Figure 13.2). This sobering chart illustrates how difficult it is for human beings to imagine their own lives over a long period of time and space. The image makes use of the conventional language of mathematic coordination systems to highlight the interaction between time and space in the context of looking at the future. The X axis marks the time, the Y axis measures the social

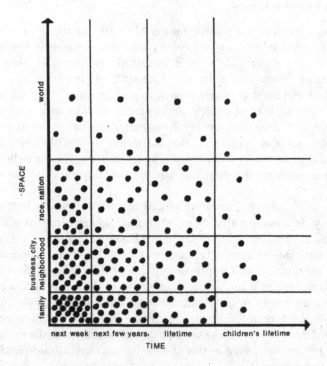

Figure 13.2 Human Perspectives. Chart taken from the *Limits to Growth* report published by the Club of Rome in 1972.

Source: Donella H. Meadows, Dennis L. Meadows, Jørgen Randers, and William W. Behrens: *Limits to Growth*, A Potomac Associates book, 1972, p.19; licence under Creative Commons BY-NC.

and geographic space. While the first segment ("family / upcoming week") is dominated by several (twenty-three) black-filled/tagged circles, the furthest area ("earth / life-span of children") is quite empty (three circle forms).

With regard to the statement made by the chart, it appears to be unimportant what exactly the individual points mean and how their number was decided upon. What is much more important is the gradual difference between empty and full. In the decreasing density of the dot pattern, it becomes apparent that people find it generally difficult in their everyday lives to imagine their own future outside of the local sphere, that is, in the meso and macro areas. The temporal distance only serves to additionally reinforce this interplay. Broadly speaking, the fact is that the further away a future scenario is from one's own standpoint, the less a person is able to imagine it. At this point, however, the methods and institutions of the forecast also change, because states and nations—just as the world community—can, as individuals, generate forecasts by other means, such as through computer simulations.

The "Human Perspectives" chart shows that people have only a limited capacity to imagine the future. For this reason, it is necessary to create professional and scientific methods of prognosis that can calculate future developments based on non-linear equations. The authors of the Club of Rome report therefore used the results of the "Human Perspectives" chart as an argument in favour of using new methods of computer simulation to overcome the lack of human beings' ability to anticipate the future. The simulation of scenarios appears to be an ideal solution—but also the "only" solution—to the challenge of being able to examine the future of complex world systems with the cool and rational eye of the scientific method and to develop action strategies based on that research.

In the triptych of scenario curves

In the 2014 IPCC summary report, the four representative scenario groups of the Representation Concentration Pathways (RCP) were reduced to two to show the maximal bandwidth in which all scenarios function. These too are the results of numerical simulations. They are, however, much more complex compared to the world model of the Club of Rome. Nevertheless, the aesthetic of the curves—especially in the early reports of the IPCC—points directly to several curve role models of the Club of Rome.

The chart contained in the SPM of the WGI report shows the spectrum of possibilities of these scenarios of the future (see Figure 13.3): the upper curve of the RCP 8.5 illustrates the steady rise in global temperatures if growth and the exploitation of the systems continue in the current unchecked manner ("rise"). The rising signature of the curve proceeds in the same manner as other growth curves relating to a "great acceleration" (such as population growth, resource consumption, etc.), that is, it follows the course of extrapolation from "business as usual". The lower line of scenario RCP 2.6 stands in contrast to the rise, it represents a much slower consumption of CO_2, which could be achieved if tough political decisions were made. The signature of this scenario is a curve, which, after a short increase, slows down and stabilises in a horizontal line of zero growth at a height of roughly 2°C ("peak and decline"). Here, the worst-case and best-case scenarios are depicted next to each other in the colours red and blue.

Much like in Hans Memling's triptych, this image makes it possible to draw a comparison between two future worlds that are the result of human behaviour on Earth (their "fault"). By extending the space of historical and present-day climate change graphically into the future, it becomes possible not only to see the present in comparison to an historical framework of "normality", but also to estimate possible futures in the same long-term time frame. This special image type thus confronts the viewer with a hybrid space that is inhabited by the past, present and future all at once, but also by different conditions of knowledge—which is why the concept of the triptych was transferred onto this image type (Walsh 2013: 179).

The first section of the curve illustrates the historical time period of the climate from 1900 to roughly 2012 in the form of a black line. This line is based on worldwide weather measurements. To estimate future temperatures,

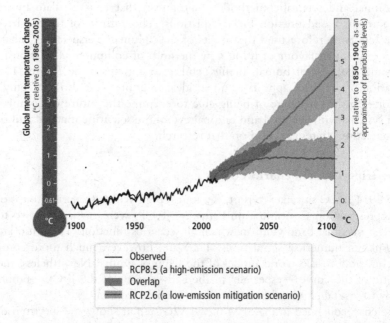

Figure 13.3 Past, present and future in one chart. What is shown are possible increases
in global average temperatures by 2100. The graph starts with a black line
based on measurements from the past, which fans out into a steeper red
line and a lower blue line with growing widths, showing the spectrum of
possible futures. These are based on the RCP scenarios 2.6 and 8.4 relative
to 1986–2005 and 1850–1900.

Source: IPCC 2013, p. 13.

the relatively solid ground of historical data is abandoned and the lines enter
the realm of numerical computer simulations: the knowledge suggested by the
lines in this section is obtained using methods of climate simulation and scenario development. The fact that this knowledge is based on other methods
is demonstrated by the blue and red areas that encompass the curve with a
spectrum of uncertainty that increases the further away the curve runs from the
present day. In turn, the thermometers on the right and left sides that frame the
Y axis—each of which marks a different zero point—reflect the current politics
involved in the atmosphere; depending on which historical reference point is
chosen, a warming of 5°C or 6°C has been determined, thus leading to different savings obligations respectively.[9] The choice of reference points determines
the level of costs or savings with which nations are called on to participate in
the process of climate protection.[10]

The concept that replaced "climate prediction" in official IPCC-speak is
"climate projection", a term originating in the field of optics and invoking
the idea of a light beam. To further examine the question of how the future is
presented in such charts, it is helpful to compare the scenario charts with the

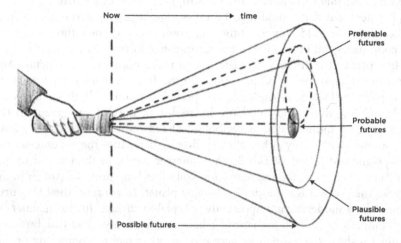

Figure 13.4 The future as a light cone. Futurology's flashlight metaphor with the distinction between possible futures, probable futures, preferable futures and plausible futures.

Source: Jessica Bland, Stian Westlake: *Don't Stop Thinking About Tomorrow: A modest defence of futurology*, Nesta, 2013, p. 9. Creative Commons BY-NC-SA.

metaphor of a flashlight, which is used in futurology to show the concept of future and examine the extent to which it can be planned (see Figure 13.4). The optical metaphor functions intuitively, seeing as it starts with the conventional idea of a gaze into the future as a path forward; it could, however, also lie in the backdrop of curves as in Figure 13.3, if these are titled as "Future Projections," for instance literally as "Future Scenario" or "Spotlight."[11]

The flashlight metaphor turns knowledge into an enlightened product of an illumination that sees the future—at least visually—as a "surprise free" zone. The cone-shaped radiance of the flashlight elucidates two elements: on the one hand, it symbolises that the only things that become visible are those things highlighted by the light of the flashlight, thus that there is an accompanying area of non-knowledge (ignorance) and that which is unimaginable. At the same time, the shape of the cone shows that knowledge about the future becomes more uncertain the further away that future is. The largest radius marks the space of all "possible" conceivable and probable futures. Three more cones are contained within this illuminated cone and are defined as the probable, plausible and preferable futures. The circle that lies somewhat shifted off from the centre—the so-called "preferable futures"—forms an intersection with the three other light cones.

The two fanned-out red and blue lines of the RCP scenarios (see Figure 13.3) can also be seen as cones of light shining into the future in either outer or inner and illuminating the realms of probability and possibility. On the one hand, in the compression of world history down to the few centimetres contained in the timeline, there lies hidden the catastrophe, on the other hand, the curve depicts no event that would reveal itself as a sudden drop or rise, that is, that would show the sudden event of the catastrophe. Indeed, the future earth heats up continually and insidiously in the surprise-free form of extrapolation.

It is part of the IPCC's mandate not to make normative statements. And yet, even though the IPCC in comparing the RCP scenarios dispenses with an explicit evaluation in the sense of "preferred futures", both scenarios are evaluated intuitively and implicitly by means of the colours given. In the multi-layered, intuitive colour symbolism of red and blue—which they have in common with many other climate charts—here, too, the worst-case scenario is marked in red (Walsh 2009; Schneider 2014). In the context of that which these charts mean, red therefore symbolises both heat and danger in the development towards a drastically changed planet. In contrast, the blue curve symbolises a moderate and apparently acceptable change. In this manner, in the comparison of both scenarios, it becomes implicitly clear that business-as-usual is the actual worst-case scenario. In other words, continuing on the current path is precisely the catastrophe that Walter Benjamin defined as being founded in technical progress:

> The concept of progress must be grounded in the idea of catastrophe. That things are *status quo* is the catastrophe. Hence [catastrophe] is not something that may happen at any time, but what in each case is given.
>
> (Benjamin 1991: 683)

Benjamin saw the ability to break through that which is given only in unpredictable states of exception: "Salvation attaches itself to the small leap in the continuous catastrophe" (Benjamin 1991: 683). Exactly where an adaptive policy or culture makes this "leap" is not evident in the framework of the chart. At this point, the question arises as to what this gaze at the future curves opens up for future scenarios beyond those abstract lines. Indeed, the charts cannot depict what the projected climate change scenarios mean for human beings and what hidden, irrational and non-measurable elements they contain. What does man know about himself in the sketch of this scenario beyond his production of greenhouse gasses if, in fact, the "*conditio humana*" is geographical (Berque), in other words, that it is climatic because that "surface entity, man" lives locally in very definite geographical and climatic conditions? Whether the lower curve—in keeping with Robert Crumb's triptych—goes hand-in-hand with "ecotopia" or with a "techno fix" is just as unclear as the effects of an "ecological disaster" on human coexistence inherent in the upper line (see Figure 13.5).

Burning worlds, or the aesthetics of heat and alarm

In addition to the scenario curves involved in possible climate futures, elliptically shaped world maps also illustrate the expected climate changes of interlinked atmosphere-ocean models in their distribution in geographical space. Maps such

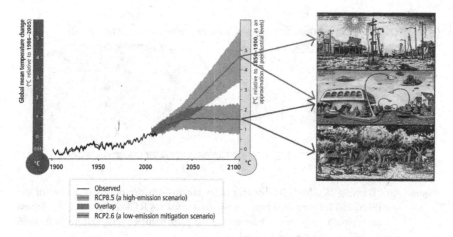

Figure 13.5 Collage of Figures 13.1 and 13.3.

as these are some of the most widely distributed scientific images of climate change. Since 2001, parallel to the reports being printed throughout in colour, they have been permeated by red, egg-shaped planets that are given a redder colour with each successive report.[12] Because they show the Blue Marble as a red planet, charts such as these were given the nickname "Burning Worlds."[13]

One chart (see Figure 13.6) taken from the SPM of the WGI in 2013, shows two such world maps with colour-filled contour lines in comparison. They demonstrate once again the future possible changes in temperatures using the two RCP 2.6 and RCP 4.5 scenarios. Here, however, the warming scenarios are not depicted in chronological order in the form of a curve (see Figure 13.3), but rather as climatography in its spatial distribution of surface temperature (2081–2100 relative to 1986–2005). The spatial distribution of thermal scenarios conveys a picture of regional climate politics to a greater degree than the abstract curve, seeing as it binds the data to specific locations rather than merely depicting the location-independent value of a radically abstract "global temperature".

The colour legend is based on so-called "false colours" (that is, colours that differ from natural colour impressions), however, their coding is adapted to the intuitive understanding of the colours blue (cool) and red (hot). The colour gradient shows a spectrum from -2 to +12°C. Bright pink follows the darkest red, and bright pink marks the largest temperature increases. The colour pink is significant, for example, hitherto unheard of record temperatures of 54°C in Australia have been mapped since summer 2013 with a new bright magenta (see weather map of Australia from 14 January 2013, Australian Bureau of Meteorology). The IPCC also sees this colour as an effective transporter of its message—to make it clear that the temperatures go beyond the existing scenario framework, the new reality was marked visually using the artificial

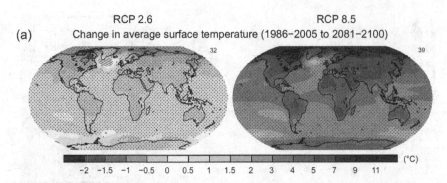

Figure 13.6 "Burning Worlds." Red world maps taken from the summary report of the IPCC 2014. Compared here are the two scenarios RCP 2.6 and RCP 8.5 (best case / worst case) for the warming of the earth's surface temperature from 1986 to 2005 in comparison to 2081 to 2100.

Source: IPCC, 2013, p. 22.

looking colour magenta. This reality reveals "the new normal" of the anomaly in a cartographic manner—an anomaly in which record-high levels have become the norm.[14]

Climate-research world maps inspire questions regarding the wider and ambiguous iconography within which these images are situated. Indeed, the red world maps achieve a new symbolic level when the visualisations that illustrate the findings of climate research make use of a mode of depiction that recalls the collective "visio type" (Pörksen 1997) of "Blue Marble." It is the image of the blue planet—that vulnerable and perfect marble—"wounded" by the colours yellow and red. The maps do not show, however, the far-off planet Mars, which is usually depicted using the colour red, instead, the maps show the earth inhabited by the viewers themselves. Covered in red, however, the maps present the visual scenario of an uninhabitable planet that knows neither life nor history.

At this point, it is possible to identify the numerous critiques of the—equally historically determined—global perspective. The godlike perspective from above makes the earth appear as an "object of contemplation, detached from the domain of lived experience" (Ingold 1995: 32). In this sense, the red-earth images also stand in the history of visualisation technologies that encouraged the separation of knowledge subject and knowledge object. The result of this separation was that the ability to develop a sense of responsibility was lost (Haraway 1989; Arendt 1992; Ingold 1995; Sloterdijk 1999). At the same time, it is precisely the global view that leaves the viewer with a feeling of the highest possible impotence and apocalyptic associations with regard to the consequences of their own behaviour. Owing to the colour and what these images mean, it is possible here for images to step out of the collective memory, images that see the future as a catastrophe, much like the image worlds of the apocalypse contained in the work of Hans Memling or

Hieronymus Bosch. However, today, these have become ideas of an "end of the world without a new beginning" (Horn 2014: 27). These two world perspectives are combined most impressively in YouTube disaster channels, which edit together catastrophe reports and footage of extreme weather events and interpret them as harbingers of the coming apocalypse, which they argue is becoming more clearly recognizable in the increase of these events.[15]

The global images generated by climate researchers can therefore be tentatively seen as cosmogram of the Anthropocene, ones that claim to tell a new chapter in the history of the cosmos and to explain its fate (Tresch 2005; Schneider 2016). The concept brings to light the similarities to other world images: the findings of climate change science contain a significant amount of that which major religions also require for their world image, such as faith, guilt and the possibility of atonement as well as the threat of an evil end to the world. Much like devotional images, they present climate change as a violent upheaval that forces human beings into a state of fatalistic powerlessness. Here, too, it's about a "future that humans make" (Horn 2014: 27). However, this prophesied future does not lie in the great hereafter, but rather in the here and now. The symptoms of a CO_2-intensive way of life are confronted in the red curves and maps. In the colourful garb of objective data visualisation, the images elucidate the tragic knowledge about the point to which "humanity" has come as a result of its globally exploitative outlook.

Blue is hope–space for opportunity and life paths

A final example comes in the form of a chart taken from the SPM of WGII (Impacts, Adaption, Vulnerability). This chart creates a fundamentally different image of the future by trying to visually cover a realm of opportunity that includes different paths for decisions and activities. In the framework of this chart, the physics of the earth is not only joined with the anthropogenic and systemic origins of the climate problem, but the solution also figuratively lies in the hands of people, that is, in the hands of policy decision makers. While the image types discussed until now also formed part of the early reports, the graphic chart with the title "Opportunity spaces and climate-resilient pathways" represents a new form of image generation that calls for a global policy approach in times of climate change (see Figure 13.7).

The chart has a three-part triptych structure. Starting from a three-coloured globe with a title that uses the cosmopolitan "we" ("our world"), a network structure originates to which four further globes are attached in the same colour scheme ("possible futures") in the right part of the image. In the depiction of the various earth versions, which are colored in red, the colour stands for human behaviour and lifestyles, including social factors that place pressure on the ecological systems ("social stressors"); they are interlinked with the biophysics of the earth (yellow), which also places pressure on the ecosystems ("biophysical stressors"), for example as a result of increasing temperatures. Within the operating area of these two systems, the ecosystems (green) find themselves symbolically

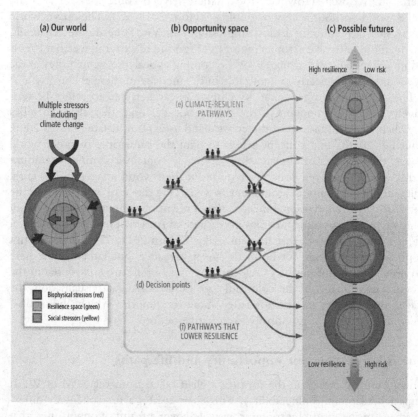

*Figure 13.*7 Our world: Opportunity spaces and possible futures. The graphic chart comes
from the Impacts, Adaption and Vulnerability Report of the IPCC from 2014.
It shows how political paths change the future of the earth. The pressure on the
different systems is symbolised by the coloured circles.

Source: IPCC 2014, p. 29.

stuck between the two grips of a pair of pliers. In the middle-area known as
"opportunity space", the nodes depict various decision options with regard to
reduced or increased resilience (the susceptibility of a system). Decisions are
made on a joint basis (grey plateaus) by a community (group of three people).

The image makes it clear that later decisions in the direction of increased
resilience will result merely in an earth with greater pressure on the durability
of its systems. Here, the image of the desired "goal earth" is summoned up to
be depicted as the ideal image in a causal relationship to the decisions made.
Unlike the metaphor of the flashlight (see Figure 13.4), the future is demonstrated in this graphic chart via the metaphor of the life path. In other words,
the future progresses differently depending on which path political decision
makers decide upon, whereby the opportunities to shape the future decrease

over time, just as in human ageing. In contrast to the previous examples, the challenge of pursuing a global policy appears feasible as a result of the radical graphic simplification of the complex decision-making path.

The semantics of colour also play an important role in this chart. The use of red, green and yellow is once again based not only on the legibility of these colours, but also on their conventional understanding. Green is the colour of environmental protection and nature, and here it stands for healthy ecosystems, while red and yellow symbolise the disturbance of this system.[16] Not only do the four possible worlds in the right section of the image get redder and redder as they go down, seeing as the human sphere increasingly restricts them, they also float against an increasingly red background that again evokes the connotation of the colour red in the force field of heat, danger and warning.

A comparison with another roadmap allows us to analyse even more precisely what the selection of the "path" metaphor means for the relationship of policymaking to science. Such a comparison was undertaken by the co-chairman of the IPCC WGIII, Ottmar Edenhofer, to illustrate the role played by climate sciences in relation to politics (see Figure 13.8).[17] It's the metaphor of a mountain hike that inspires a number of possible interpretations.

The image demonstrates that science offers feasible paths and options for policy makers. In other words, much like a map of knowledge, science

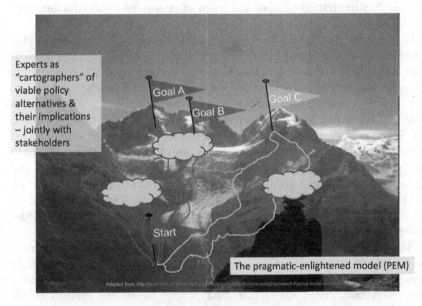

Figure 13.8 Science as a cartograph of viable policy options. Illustration in the form of a hiking path, PowerPoint slide by Ottmar Edenhofer, 2012.

Source: PowerPoint slide by Ottmar Edenhofer and Martin Kowarsch, 2015: Public Policy Assessments: Being Policy Relevant without being Policy-Prescriptive. Experiences With The IPCC. Published on www.pik-potsdam.de (5/10/2016). Background photograph: © Willy Gisler.

cartographs all possible paths to illustrate the choices policymakers can take. The actual path chosen by policymakers viewing this map depends considerably on the goals that are agreed upon politically. The image of the map as a metaphor for the future illustrates that each path—whether it's along a slope, over a glacier or a field of snow—will be strenuous. However, the image also suggests that science—in the role of an experienced and reliable mountain guide (in book form)—knows the terrain and can read the signs. In other words, science is an entity that knows when the weather threatens to change and when blue skies will turn into torrential streams, at which point the path over the ridge becomes impossible and the only thing left to do is descend into the valley.

Data images as associative spaces

To sum up, it is beneficial to compare the different future scenario types. Much like the folding panels of Hans Memling's altarpiece, in the case of the scientific visions of the future, different conceptional areas in alternative world paths open up next to one another. These worlds depend on the actions of humanity, and they suggest either recovery or the end of the world: in addition, they are divided temporally, seeing as they usually juxtapose extreme best-case and worst-case scenarios. The scientific images make use of this image structure because they illustrate the future as a realm of opportunity that contains alternative realities.

In particular, in the reports of the IPCC, red maps and curves show what is known about the earth's future climate based on current scientific knowledge. The analysis of the examples is designed to show how, when examining scientific images, the possible futures of the earth can be shown *globally* and how already existing future narratives are evoked. These are interpreted as collective-imaginary revenants into these images—much like in a Rorschach test. When such charts manage to enter the non-scientific realm—often robbed of their scientific explanations—it becomes almost impossible to limit them to their otherwise professional manner of interpretation of pure denotation. The "culture of prediction" from which these images arise therefore cannot be said to be purely an internal feature of the atmospheric sciences; rather, it is a culture resolutely situated within, and in constant dialogue with, much wider modes of comprehending and engaging with the future.

At the same time, it remains unclear in the rational images of the future—the ones that are simulated in the young regime of anticipation and of strategic future management using scenario technology (Schmidt-Gernig 2002)—what these projections mean for cultures and people. While the creators of the rational images of the future want to draw images of possible future presents, they set off among viewers present imaginations of futures. Their rationality, as it is shown in particular in the interpretation of these images in the scheme of the apocalypse, leads to emotional interpretations. This works perhaps that much better because the colourful maps and curves remain quite featureless in their frightening depictions of futures up until 2100. The abstract emptiness of these future spaces can therefore be filled that much more successfully in the mind's eye of the recipient using already known narratives and collective phantasms about the end of the world. In fact, these curves and maps depicting possible climate futures ultimately surpass the imaginative capacity with regard to humanity's destructive effects on the earth. Lorraine Daston uses the terms "horror" and "terror" to describe that which is triggered when feelings can't react (Daston 1998). Where the individual's ability to imagine the future fails, major cultural narratives of trans-personal scenarios push their way into the foreground, every individual has such scenarios and they are dependent on education, influences and origins.[18] This way, all possible forms of "imagery" produced by a culture can be linked by association to the dry skeleton of charts, which are concerned with the entirety of the world; in other words, they trace a cosmogram of the earth.

The cultural perspective on these images is particularly evident in the "Burning Worlds" nickname given to the temperature maps. The crux here is that the global cartographies of more dangerous climate futures become images of fear, and these images wear out in the course of media reproduction. At this point, the question arises as to whether images such as these—the aesthetics of such images—might actually lead from knowledge to action, or whether they, like Medusa, can only lead to the freezing and paralysis of the gaze. The example of a future of the earth using the metaphor of the path—which, in contrast to curves and maps, shows people who are shaping and designing this future—speaks a different language, one that appeals directly to the behaviours and decision-making ability of political levels. In other words, it shows the consequence of the strategic decision points that result from the scenario technique.

Thus, it would appear that the notions of the future that come from the scientific images on climate change are, as such, highly limited in their repertoire. Schematically, they consist of the three futures that Robert Crumb created in his comic more than forty years ago: the vision of a technical healing, of an eco-utopia and of an ecological catastrophe, the latter often presented as a ruined earth after the extinction of humanity. The progress-optimistic interpretation reflects the extrapolated vision of a necessary "techno fix"; the apocalyptical interpretation reflects the vision of an inevitable end of the world and bears the character of a catastrophe with the threatening future as a punishment for accumulated debt; the third and least widespread vision—and the one unimaginable for many people—is that of a retrograde ecotopia.[19]

In my opinion, if we are to expand our ability to imagine the future so as to be able to conceive possible worlds beyond the conventional interpretations alluded to in the associative data spaces, then the data involved in climate research must begin to tell stories in other media and genres as well. Philosopher Günther Anders called for a stretch of the imagination such as this when he—in the face of the equally inconceivable possibility of a nuclear strike—sought to reduce the "Promethean gap" between technology and feeling by means of moral stretching exercises. Much like those based on probability theory, such narratives are equally based on fictional premises (Esposito 2007: 55). For example, art, films and literature offer an imaginative instrument with which the opportunity spaces and "story lines" of a future could be further told and explored. The data begin to tell stories that go far beyond that which they describe in the abstract space of numbers. At the centre is the idea of how a changed climate will also change how humanity will live together. The current stories and visions contained in films, literature and computer games offer diverse material for the imagination that can fill out the abstract spaces of climatological maps and curves with end-of-days narratives or alternative world designs—and thus, in this way, have an effect on the shape of the future. Today's Memlings and Boschs are authors such as J. G. Ballard, Barbara Kingsolver, T. Coraghessan Boyle, film directors such as Bong Joon-ho, Tim Fehlbaum and Roland Emmerich, computer game authors such as those of *Anno 2070* and photographers such as Robert Graves and Didier Madoc, who outline futures of cities and societies, or artists like Eve Mosher, who makes predictions about where waters will rise due to climate change with her *High WaterLine* performances. These artists also create storylines and use fictional and pictorial means to simulate possible climatic upheavals such as those of a RCP 4.5 earth. In doing so, they make a world imaginable that shines only dimly through the skeleton of the red scenarios.

Notes

1 *CoEvolutionary Quartlerly* was the successor to the *Whole Earth Catalogue* and was also published by Stewart Brand. It was in print from 1974–1985.

2 The original can be found in Danzig. There is also a copy in the Gemäldegalerie in Berlin. The image was created in Brügge on commission by the chairman of the Medici bank for a church in Florence. It never arrived there, however, due to a robbery during transport via sea.

3 The image of an apocalyptic end time that will ultimately impact all people is a common concept in monotheistic religions, i.e. Islam, Christianity and Judaism. Historically it goes as far back as the Babylonian and Egyptian notions of the hereafter. For devout Christians—and especially in the Middle Ages, but also long afterwards—the final judgement possessed a validity that outshone all other concepts of the future: it involved the belief that the final judgement was imminent and it was designed to inspire followers to live a good and righteous life.

4 This issue was studied more intensively with different methods including a survey by Schneider and Nocke (forthcoming 2017).

5 In the Climate Change Synthesis Report for Policy Makers published in 2014, which brings together all reports, there are fourteen charts (without tables), ten of which show possible futures and future risks. In the Synthesis Report from 2007, six of nine images were images of the future: in 2001, it was ten out of twelve. In the first report from 1990, there was only one image (in the conclusion) that depicted a future scenario. In the associated summaries from 1992 issued by the three Working Groups, the following was the case: in the summary of WG I, nine of fifteen images that showed possible futures; in the summary of WG II, there were no illustrations; and in the summarizing report of the WG III, there were three illustrations, of which one showed futures. The earlier reports were black and white. They have been in colour since 2001.

6 Even if the IPCC stipulates that the captions must remain attached to the images, in practice, these requirements are rarely met. Images function in a "immutable mobile"

manner (Bruno Latour) as transit media that are stable in their form; they can be easily copied and distributed, which is why they are still more important than the interactive graphics on climate change that now exist.

7 The definition of these concepts is clearly distinguished in the IPCC report. A "climate prediction" is the result of processes such as "projections" and "scenarios". Scenarios are the simplified models of climatological relations with which simulations are carried out so as to examine the consequences of the anthropogenic climate change. The projections, in turn, refer to assumptions about possible future anthropogenic emissions and their influences. The concept of "climate projection" has replaced "climate prediction" so as to clarify the status of probability and vague knowledge (Gramelsberger 2012).

8 Climate change did not yet play a role in the publication, although a forecast with regard to increases in CO_2 in the atmosphere was made. For the field of ecology, the focus was more on questions relating to nuclear waste, resource consumption, the depletion of fish stocks in oceans and food production. Several scenarios were tested, such as unlimited population growth and industry versus the stabilization of population figures, etc., whereby one result of the study was the suggestion that the balance (equilibrium) and stabilization of the world model was a desirable goal. Ways to solve the problem were offered.

9 The reference point of warming (0°C) is set either in relation to the pre-industrial phase from 1850 to 1900 (right) or to the phase from 1986 to 2005 (left). In the first case, there is a warming up to 6°C (upper red edge), in the second case a warming of only 5°C. The question of this historical reference point is essential for the policies undertaken with regard to the 2°C goal, which is itself politically motivated; this is why they are shown in the image as well.

10 This is why the curve is found next to an other chart that shows the slim lines of the scenarios in relation to the rising risks and hazards. The chart was formally called "Reasons for Concern" and was given the nickname "Burning Embers" as a result of its firery red colours (Mahony 2014).

11 The word-field of the term "projection" is varied depending on the area of application. Projection means the depiction of an object or its image on an image level through a lens or a lens system. In psychology, in turn, it refers to the unconscious transfer of one's own desires, feelings and ideas onto other persons or objects. In statistics, it refers to a process designed to derive a third value out of two statistical values.

12 The projection type is called Mollweide Projection. In contrast to the Mercator Projection, the Mollweide Projection allows for a relatively equal-area depiction of the earth.

13 'Burning world' is a common category in stock image data bases, like Getty, to illustrate global warming. The double meaning of 'burning' as in heat and burning fossils are found in book titles such as *How to stop the planet from burning*, and in science fiction series taking place in a post-climate-catastrophic world such as *Aviator: The Burning World* by Gareth Renowden.

14 The Secretary General of the United Nations, Ban Ki-moon, who took part in the conference, took this natural catastrophe as an opportunity to classify the event in the context of a new normality when he said "that climate change is making extreme weather events the new normal" (Wade 215).

15 See the YouTube channels *Angel Apocalypse* and *Signs of 2014*. The *Angel Apocalypse* channel advertises using the following sentence: "Be witness to how the earth is crumbling into pieces by devastating tornadoes, erupting volcanoes, massive earthquakes and never-ending floods. [...] Be sure that's not the end of the world which I'm talking about, it's the end of this age before our great savior Jesus returns" (*Angel Apocalypse* 2014).

16 The fact that the background color changed upon completion of the report from a light orange to a light, more hopeful blue demonstrates once again the multifaceted connotations of blue and red in this field.

17 Ottmar Edenhofer is co-chairman of the WGIII of the IPCC, Director of the Mercator Research Institute on Global Commons and Climate Change (MCC), and Deputy Director and Head Economist at the Potsdam Institute for Climate Change Research.

18 Drawing on Sigmund Freud, such graphics have the possibility of an "overdetermination by means of the system of each dominant collective symbolism" in the sense of a polyphony or a surplus (Link 2001: 115).
19 What people hope from the future was surveyed in a Norwegian study recently using the Robert Crumb comic. Elementary school children were asked for their thoughts about the future. The results showed that hardly any of those surveyed could imagine the utopian alternative, while the majority expected the worst-case scenario of the ecological catastrophe and hoped for the technical solution (Fløttum et al. 2016).

References

Adams, V., M. Murphy, and A. E. Clarke (2009) "Anticipation: Technoscience, life, affect, temporality," *Subjectivity* 28, pp. 246–265.

Arendt, H. (1992) *Vita activa oder Vom tätigen Leben* (The Human Condition), Munich: Beck.

Benjamin, W. (1991) *Gesammelte Schriften*, (vol. 1), Frankfurt: Suhrkamp (translation taken from W. Benjamin (2003) *Selected Writings: 1938–1940*, Cambridge, MA: The Belknap Press of Harvard University Press).

Coevolution quarterly (1981) *Whole Earth Catalog*, Sausalito, CA: Point.

Daston, L. and K. Park (1998) *Wonders and the Order of Nature: 1150–1750*, New York, NY: Zone Books.

Edenhofer, O., Minx, J. (2014) "Mapmakers and navigators, facts and values," *Science* 345(6192), pp. 37–38.

Edenhofer, O. and M. Kowarsch (2015) "Cartography of pathways: A new model for environmental policy assessments," *Environmental Science & Policy* 51, pp. 56–64.

Edenhofer, O. and M. Kowarsch (2015) "Ausbruch aus dem stahlharten Gehäuse der Hörigkeit: ein neues Modell wissenschaftlicher Politikberatung," P. Weingart and G. G. Wagner (eds.) *Wissenschaftliche Politikberatung im Praxistest*, Velbrück Wissenschaft Verlag, pp. 83–106.

Esposito, E. (2007) *Die Fiktion der wahrscheinlichen Realität*, Frankfurt, Germany: Suhrkamp.

Fløttum, K., T. Dahl, and V. Rivenes (2016) "Young Norwegians and their views on climate change and the future: Findings from a climate concerned and oil rich nation," *Journal of Youth Studies* 19(8), pp. 1128–1143.

Gramelsberger, G. (2013) "Intertextualität und Projektionspotential von Klimasimulationen," D. Weidner and S. Willer (eds.) *Prophetie und Prognostik, Verfügungen über Zukunft in Wissenschaften, Relationen und Künsten*, Munich, Germany: Fink, pp. 209–225.

Haraway, D. (1989) *Primate Visions: Gender, Race, and Nature in the World of Modern Science*, New York, NY: Routledge.

Horn, E. (2014) *Zukunft als Katastrophe* (Future as Catastrophe), Frankfurt, Germany: Suhrkamp.

Hulme, M. (2009) *Why We Disagree About Climate Change*. Cambridge, MA: Cambridge University Press.

Ingold, T. (1995) "Globes and spheres," K. Milton (ed.) *Environmentalism. The View from Anthropology*, London: Routledge, pp. 31–42.

IPCC (2013) "Summary for policymakers," T. F. Stocker, D. Qin, G.-K. Plattner, et al. (eds.) *Climate Change 2013: The Physical Science Basis. Contribution of Working Group I to the Fifth Assessment Report of the Intergovernmental Panel on Climate Change*, Cambridge, MA: Cambridge University Press, pp. 1–27.

IPCC (2014) "Summary for policymakers," C. B. Field, V. R. Barros, D. J. Dokken, et al. (eds.) *Climate Change 2014: Impacts, Adaptation, and Vulnerability. Part A: Global and Sectoral Aspects. Contribution of Working Group II to the Fifth Assessment Report of the Intergovernmental Panel on Climate Change*, Cambridge, MA: Cambridge University Press, Cambridge, pp. 1–32.

Link, J. (2002) "Das ‚normalistische' Subjekt und seine Kurven. Zur symbolischen Visualisierung orientierender Daten," D. Gugerli and B. Orland (eds.) *Ganz normale Bilder. Historische Beiträge zur visuellen Herstellung von Selbstverständlichkeit*, Zurich: Chronos, pp. 107–128.

Mahony, M. (2014) "The color of risk: Expert judgment and diagrammatic reasoning in the IPCC's 'Burning Embers'," T. Nocke and B. Schneider (eds.) *Image Politics of Climate Change, Visualizations, Imaginations, Documentations*, Bielefeld, Germany: transcript, pp. 105–124.

Manzo, K. (2010) "Imaging vulnerability: The iconography of climate change," *Area* 42(1), pp. 96–107.

Meadows, D. H. (1972) *The Limits to Growth. A Report for the Club of Rome's Project on the Predicament of Mankind*, New York, NY: Universe Books.

Nesta Organisation (2013) *Don't Stop Thinking About Tomorrow: A Modest Defence of Futurology*, available from https://www.nesta.org.uk/sites/default/files/dont_stop_thinking_about_tomorrow.pdf, accessed March 2015.

Pörksen, U. (1997) *Weltmarkt der Bilder: Eine Philosophie der Visiotype*, Suttgart, Germany: Klett-Cotta.

Rosentrater, L. D. (2010) "Representing and using scenarios for responding to climate change," *WIREs Climate Change* 1, pp. 253–259.

Schmidt-Gernig, A. (2002) "Ansichten einer zukünftigen "Weltgesellschaft". Westliche Zukunftsforschung der 60er und 70er Jahre als Beispiel einer transnationalen Expertenöffentlichkeit," H. Kaelble, M. Kirsch and A. Schmidt-Gernig (eds.) *Transnationale Öffentlichkeiten und Identitäten im 20. Jahrhundert*, Frankfurt, Germany: Campus, pp. 393–421.

Schneider, B. (2014) "Red futures. The colour red in scientific imagery of climate change," M. Juneja and G.J. Schenk (eds.) *Disaster as Image: Iconographies and Media Strategies across Europe and Asia*, Regensburg, Germany: Schnell & Steiner, pp. 183–193.

Schneider, B. (2016) "Burning worlds of cartography. A critical approach to climate cosmograms of the Anthropocene," *Geo: Geography and Environment* 3(2). DOI:10.1002/geo2.27

Schneider, B. and T. Nocke (forthcoming 2017) "The feeling of red and blue – a constructive critique of color mapping in visual climate change communication," W. Leal Filho (ed.) *Handbook of Climate Change Communication*, Cham, Switzerland: Springer.

Sloterdijk, P. (1999) *Globen*, Frankfurt, Germany: Suhrkamp.

Tresch, J. (2005) "Cosmogram. Interview with Jean-Christophe Royoux," M. Ohanian and J.-C. Royoux (eds.) *Cosmograms*, New York, Berlin: Lukas and Sternberg, pp. 67–76.

Wade, M. (2015) "Extreme weather the new normal in Australia's disaster-prone neighbourhood," *The Sydney Morning Herald*, 16 March 2015.

Walsh, L (2013) *Scientists as Prophets: A Rhetorical Genealogy*, Oxford, UK: Oxford University Press.

Walsh, L. (2009) "Visual strategies to integrate ethos across the 'is/ought' divide in the IPCC's Climate Change 2007: Summary for policy makers," *Poroi An Interdisciplinary Journal of Rethorocal Analysis and Invention* 6(2), pp. 33–61.

Walter, F. (2010) *Katastrophen. Eine Kulturgeschichte vom 16. bis ins 21. Jahrhundert (Catastrophes: A Cultural History from the 16th to the 20th Century)*, Stuttgart, Germany: Reclam.

Wildavsky, A. (1982) *Risk and Culture: An Essay on the Selection of Technical and Environmental Dangers*, Berkeley, CA: University of California Press.

Index

Printed in the United States
By Bookmasters